Cambridge International AS & A Level

Mathematics
Probability & Statistics 2

Sophie Goldie
Series editor: Roger Porkess

Questions from the Cambridge International AS & A Level Mathematics papers are reproduced by permission of Cambridge Assessment International Education. Unless otherwise acknowledged, the questions, example answers, and comments that appear in this book were written by the authors. Cambridge Assessment International Education bears no responsibility for the example answers to questions taken from its past question papers which are contained in this publication.

®IGCSE is a registered trademark.

The publishers would like to thank the following who have given permission to reproduce photographs in this book:

Photo credits: page 1 *top* © Clive Chilvers / Shutterstock; *bottom* © Charlie Edwards / Getty Images; page 11 © Eric Gevaert / Shutterstock; page 14 © Ingram Publishing Limited / Ingram Image Library 500-Animals; page 34 © Adul10 / Shutterstock; page 45 © Claudia Paulussen / Fotolia.com; page 56 *top* © mastersky / stock.adobe.com; *bottom* © Stuart Miles / Fotolia.com; page 84 © Ingram Publishing Limited; page 92 © Peter Titmuss / Alamy Stock Photo; page 97 © Monkey Business / Fotolia; page 113 *top* © gunnar3000 / stock.adobe.com; *bottom* © Kevin Peterson / Photodisc / Getty Images

We are grateful to Anthony Cutler who has given permission to use his quotation on page 56.

Orders: please contact Bookpoint Ltd, 130 Park Drive, Milton Park, Abingdon, Oxon OX14 4SE. Telephone: (44) 01235 827720. Fax: (44) 01235 400401. Email education@bookpoint.co.uk. Lines are open from 9 a.m. to 5 p.m. Monday to Saturday, with a 24-hour message answering service. You can also order through our website: www.hoddereducation.com

Much of the material in this book was published originally as part of the MEI Structured Mathematics series. It has been carefully adapted for the Cambridge International AS & A Level Mathematics syllabus. The original MEI author team for Statistics comprised Alec Cryer, Michael Davies, Anthony Eccles, Bob Francis, Gerald Goddall, Alan Graham, Nigel Green, Liam Hennessey, Roger Porkess and Charlie Stripp.

First published in 2018 by
Hodder Education,
an Hachette UK company,
Carmelite House,
50 Victoria Embankment,
London EC4Y 0DZ

Impression number 5 4 3 2 1

Year 2022 2021 2020 2019 2018

Cover photo by © Shutterstock/osh
Illustrations by Pantek Media, Maidstone, Kent & Integra Software Services Pvt. Ltd, Pondicherry, India
Typeset in Bembo Std 11/13pt by Integra Software Services Pvt. Ltd, Pondicherry, India
Printed in Italy

A catalogue record for this title is available from the British Library.

ISBN 97815104 2177 6

Contents

Introduction

This is one of a series of five books supporting the Cambridge International AS & A Level Mathematics 9709 syllabus for examination from 2020. The series then continues with four more books supporting Cambridge International AS & A Level Further Mathematics 9231. It follows on from *Probability & Statistics 1*. The six chapters in this book cover the probability and statistics required for the Paper 6 examination. This part of the series also contains two books for pure mathematics and one book for mechanics.

These books are based on the highly successful series for the Mathematics in Education and Industry (MEI) syllabus in the UK but they have been redesigned and revised for Cambridge International students; where appropriate, new material has been written and the exercises contain many past Cambridge International examination questions. An overview of the units making up the Cambridge International syllabus is given in the following pages.

Throughout the series, the emphasis is on understanding the mathematics as well as routine calculations. The various exercises provide plenty of scope for practising basic techniques; they also contain many typical examination-style questions.

The original MEI author team would like to thank Sophie Goldie who has carried out the extensive task of presenting their work in a suitable form for Cambridge International students and for her many original contributions. They would also like to thank Cambridge Assessment International Education for its detailed advice in preparing the books and for permission to use many past examination questions.

Roger Porkess

Series editor

How to use this book

The structure of the book

This book has been endorsed by Cambridge Assessment International Education. It is listed as an endorsed textbook for students taking the Cambridge International AS & A Level Mathematics 9709 syllabus. The Probability & Statistics 2 syllabus content is covered comprehensively and is presented across six chapters, offering a structured route through the course.

The book is written on the assumption that you have covered and understood the content of the Cambridge IGCSE® Mathematics 0580 (Extended curriculum) or Cambridge O Level Mathematics 4024/4029 syllabus. The following icon is used to indicate material that is not directly on the syllabus.

e There are places where the book goes beyond the requirements of the syllabus to show how the ideas can be taken further or where fundamental underpinning work is explored. Such work is marked as **extension**.

Each chapter is broken down into several sections, with each section covering a single topic. Topics are introduced through **explanations**, with **key terms** picked out in red. These are reinforced with plentiful **worked examples**, punctuated with commentary, to demonstrate methods and illustrate application of the mathematics under discussion.

Regular **exercises** allow you to apply what you have learned. They offer a large variety of practice and higher-order question types that map to the key concepts of the Cambridge International syllabus. Look out for the following icons.

PS **Problem-solving questions** will help you to develop the ability to analyse problems, recognise how to represent different situations mathematically, identify and interpret relevant information, and select appropriate methods.

M **Modelling questions** provide you with an introduction to the important skill of mathematical modelling. In this, you take an everyday or workplace situation, or one that arises in your other subjects, and present it in a form that allows you to apply mathematics to it.

CP **Communication and proof questions** encourage you to become a more fluent mathematician, giving you scope to communicate your work with clear, logical arguments and to justify your results.

Exercises also include questions from real Cambridge Assessment International Education past papers, so that you can become familiar with the types of questions you are likely to meet in formal assessments.

IGCSE® is a registered trademark.

Answers to exercise questions, excluding long explanations and proofs, are included in the back of the book, so you can check your work. It is important, however, that you have a go at answering the questions before looking up the answers if you are to understand the mathematics fully.

In addition to the exercises, a range of additional features is included to enhance your learning.

ACTIVITY

Activities invite you to do some work for yourself, typically to introduce you to ideas that are then going to be taken further. In some places, activities are also used to follow up work that has just been covered.

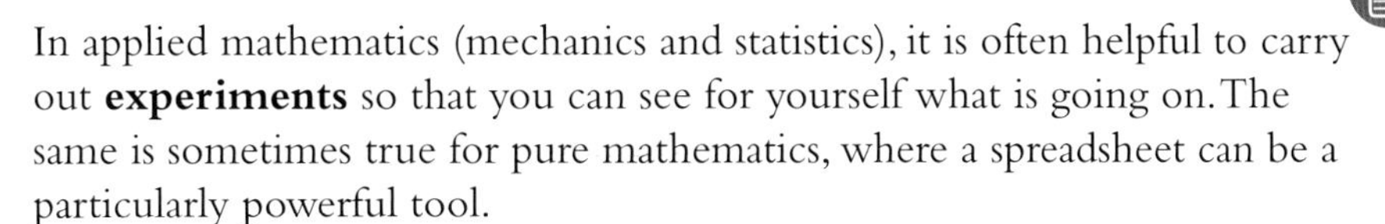

EXPERIMENT

In applied mathematics (mechanics and statistics), it is often helpful to carry out **experiments** so that you can see for yourself what is going on. The same is sometimes true for pure mathematics, where a spreadsheet can be a particularly powerful tool.

Other helpful features include the following.

? This symbol highlights points it will benefit you to **discuss** with your teacher or fellow students, to encourage deeper exploration and mathematical communication. If you are working on your own, there are answers in the back of the book.

! This is a **warning** sign. It is used where a common mistake, misunderstanding or tricky point is being described to prevent you from making the same error.

A variety of notes is included to offer advice or spark your interest:

Note

Notes expand on the topic under consideration and explore the deeper lessons that emerge from what has just been done.

Historical note

Historical notes offer interesting background information about famous mathematicians or results to engage you in this fascinating field.

Finally, each chapter ends with the **key points** covered, plus a list of the **learning outcomes** that summarise what you have learned in a form that is closely related to the syllabus.

Digital support

Comprehensive online support for this book, including further questions, is available by subscription to MEI's Integral® online teaching and learning platform for AS & A Level Mathematics and Further Mathematics, integralmaths.org. This online platform provides extensive, high-quality resources, including printable materials, innovative interactive activities, and formative and summative assessments. Our eTextbooks link seamlessly with Integral, allowing you to move with ease between corresponding topics in the eTextbooks and Integral.

Additional support

The **Question & Workbooks** provide additional practice for students. These write-in workbooks are designed to be used throughout the course.

The **Study and Revision Guides** provide further practice for students as they prepare for their examinations.

These supporting resources and MEI's Integral® material have not been through the Cambridge International endorsement process.

The Cambridge International AS & A Level Mathematics 9709 syllabus

The syllabus content is assessed over six examination papers.

Paper 1: Pure Mathematics 1	Paper 4: Mechanics
• 1 hour 50 minutes • 60% of the AS Level; 30% of the A Level • Compulsory for AS and A Level	• 1 hour 15 minutes • 40% of the AS Level; 20% of the A Level • Offered as part of AS or A Level
Paper 2: Pure Mathematics 2	**Paper 5: Probability & Statistics 1**
• 1 hour 15 minutes • 40% of the AS Level • Offered only as part of AS Level; not a route to A Level	• 1 hour 15 minutes • 40% of the AS Level; 20% of the A Level • Compulsory for A Level
Paper 3: Pure Mathematics 3	**Paper 6: Probability & Statistics 2**
• 1 hour 50 minutes • 30% of the A Level • Compulsory for A Level; not a route to AS Level	• 1 hour 15 minutes • 20% of the A Level • Offered only as part of A Level; not a route to AS Level

The following diagram illustrates the permitted combinations for AS Level and A Level.

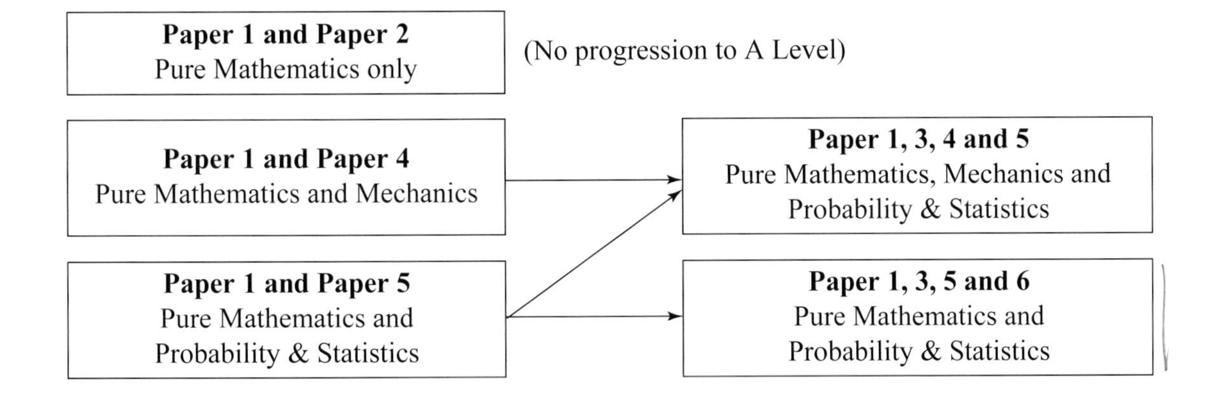

Prior knowledge

Knowledge of the content of the Cambridge IGCSE® Mathematics 0580 (Extended curriculum), or Cambridge O Level 4024/4029, is assumed. Learners should be familiar with scientific notation for compound units, e.g. $5\,\mathrm{m\,s^{-1}}$ for 5 metres per second.

In addition, learners should:

- be able to carry out simple manipulation of surds (e.g. expressing $\sqrt{12}$ as $2\sqrt{3}$ and $\frac{6}{\sqrt{2}}$ as $3\sqrt{2}$)
- know the shapes of graphs of the form $y = kx^n$, where k is a constant and n is an integer (positive or negative) or $\pm\frac{1}{2}$.

Knowledge of the content of Paper 5: Probability & Statistics 1 is assumed, and learners may be required to demonstrate such knowledge in answering questions. Knowledge of calculus within the content for Paper 3: Pure Mathematics 3 will also be assumed.

Command words

The table below includes command words used in the assessment for this syllabus. The use of the command word will relate to the subject context.

Command word	What it means
Calculate	work out from given facts, figures or information
Describe	state the points of a topic / give characteristics and main features
Determine	establish with certainty
Evaluate	judge or calculate the quality, importance, amount, or value of something
Explain	set out purposes or reasons / make the relationships between things evident / provide why and/or how and support with relevant evidence
Identify	name/select/recognise
Justify	support a case with evidence/argument
Show (that)	provide structured evidence that leads to a given result
Sketch	make a simple freehand drawing showing the key features
State	express in clear terms
Verify	confirm a given statement/result is true

Key concepts

Key concepts are essential ideas that help students develop a deep understanding of mathematics.

The key concepts are:

Problem solving

Mathematics is fundamentally problem solving and representing systems and models in different ways. These include:

- Algebra: this is an essential tool which supports and expresses mathematical reasoning and provides a means to generalise across a number of contexts.
- Geometrical techniques: algebraic representations also describe a spatial relationship, which gives us a new way to understand a situation.
- Calculus: this is a fundamental element which describes change in dynamic situations and underlines the links between functions and graphs.
- Mechanical models: these explain and predict how particles and objects move or remain stable under the influence of forces.
- Statistical methods: these are used to quantify and model aspects of the world around us. Probability theory predicts how chance events might proceed, and whether assumptions about chance are justified by evidence.

Communication

Mathematical proof and reasoning is expressed using algebra and notation so that others can follow each line of reasoning and confirm its completeness and accuracy. Mathematical notation is universal. Each solution is structured, but proof and problem solving also invite creative and original thinking.

Mathematical modelling

Mathematical modelling can be applied to many different situations and problems, leading to predictions and solutions. A variety of mathematical content areas and techniques may be required to create the model. Once the model has been created and applied, the results can be interpreted to give predictions and information about the real world.

These key concepts are reinforced in the different question types included in this book: **Problem-solving**, **Communication and proof**, and **Modelling**.

1 Sampling

If you wish to learn swimming you have to go into the water.
George Pólya (1887–1985)

Politics Now

Independent set to become member of parliament

Next week's local election looks set to produce the first independent member of parliament for many years, according to an opinion poll conducted by the 'Politics Now' team.

When 30 potential voters were asked who they thought would make the best member of parliament, 12 opted for Independent candidate Mrs Chalashika. The other three candidates attracted between three and nine votes.

Mrs Grace Chalashika is taking the polls by storm.

> Assuming that the figures quoted in the article are true, does this really mean that Independent Mrs Chalashika will be elected to Parliament next week?

Only time will tell, but meanwhile the newspaper report raises a number of questions that should make you suspicious of its conclusion.

Was the sample large enough? Thirty seems a very small number.

Were those interviewed asked the right question? They were asked who they thought would make the best member of parliament, not who they intended to vote for.

How was the sample selected? Was it representative of the whole electorate?

Before addressing these questions, you will find it helpful to be familiar with the language and notation associated with sampling.

1.1 Terms and notation

'Politics Now' took a sample of size 30. Taking samples and interpreting them is an essential part of statistics. The populations in which you are interested are often so large that it would be quite impractical to use every item; the electorate for that area might well number 70 000.

A **sample** provides a set of data values of a random variable, drawn from all such possible values, the **parent population**. The parent population can be finite, such as all professional footballers, or infinite, such as the points where a dart can land on a dart board.

A representation of the items available to be sampled is called the **sampling frame**. This could, for example, be a list of the sheep in a flock, a map marked with a grid or an electoral register. In many situations no sampling frame exists nor is it possible to devise one, for example for the cod in the North Atlantic. The proportion of the available items that are actually sampled is called the **sampling fraction**.

A parent population, often just called the **population**, is described in terms of its **parameters**, such as its mean, μ, and variance, σ^2. By convention Greek letters are used to denote these parameters.

A value derived from a sample is written in Roman letters: mean, $\bar{x}$, variance, s^2, etc. Such a number is the value of a **sample statistic** (or just **statistic**). When sample statistics are used to estimate the parent population parameters they are called **estimates**.

Thus if you take a random sample in which the mean is $\bar{x}$, you can use $\bar{x}$ to estimate the parent mean, μ. If in a particular sample $\bar{x} = 23.4$, then you can use 23.4 as an estimate of the population mean. The true value of μ will generally be somewhat different from your estimated value.

Upper case letters, X, Y, etc., are used to represent the random variables, and lower case letters, x, y, etc., to denote particular values of them. In the example of 'Politics Now''s survey of voters, you could define X to be the percentage of voters, in a sample of size 30, showing support for Mrs Chalashika. The particular value from this sample, x, is $\left(\frac{12}{30}\right) \times 100 = 40\%$.

1.2 Sampling

There are essentially two reasons why you might wish to take a sample:

- » to estimate the values of the parameters of the parent population
- » to conduct a hypothesis test.

There are many ways you can interpret data. First you will consider how sample data are collected and the steps you can take to ensure their quality.

An estimate of a parameter derived from sample data will in general differ from its true value. The difference is called the **sampling error**. To reduce the sampling error, you want your sample to be as representative of the parent population as you can make it. This, however, may be easier said than done.

Here are a number of questions that you should ask yourself when about to take a sample.

1 Are the data relevant?

It is a common mistake to replace what you need to measure by something else for which data are more easily obtained.

You must ensure that your data are relevant, giving values of whatever it is that you really want to measure. This was clearly not the case in the example of the 'Politics Now' survey, where the question people were asked, 'Who would make the best member of parliament?', was not the one whose answer was required. The question should have been 'Which person do you intend to vote for?'

2 Are the data likely to be biased?

Bias is a systematic error. If, for example, you wished to estimate the mean time of young women running 100 metres and did so by timing the members of a hockey team over that distance, your result would be biased. The hockey players would be fitter and more athletic than most young women and so your estimate for the time would be too low.

You must try to avoid bias in the selection of your sample.

3 Does the method of collection distort the data?

The process of collecting data must not interfere with the data. It is, for example, very easy when designing a questionnaire to frame questions in such a way as to lead people into making certain responses. 'Are you a law-abiding citizen?' and 'Do you consider your driving to be above average?' are both questions inviting the answer 'Yes'.

In the case of collecting information on voting intentions another problem arises. Where people put the cross on their ballot papers is secret and so people are being asked to give away private information. There may well be those who find this offensive and react by deliberately giving false answers.

People often give the answer they think the questioner wants to receive.

4 Is the right person collecting the data?

Bias can be introduced by the choice of those taking the sample. For example, a school's authorities want to estimate the proportion of the students who smoke, which is against the school rules. Each class teacher is told to ask five students whether they smoke. Almost certainly some smokers will say 'No' to their teacher for fear of getting into trouble, even though they might say 'Yes' to a different person.

5 Is the sample large enough?

The sample must be sufficiently large for the results to have some meaning. In this case the intention was to look for differences of support between the four candidates and for that a sample of 30 is totally inadequate. For opinion polls, a sample size of about 1000 is common.

The sample size depends on the precision required in the results. For example, in the opinion polls for elections a much larger sample is required if you want the estimate to be reliable to within 1% than if 5% will do.

6 Is the sampling procedure appropriate in the circumstances?

The method of choosing the sample must be appropriate. Suppose, for example, you were carrying out the survey of people's voting intentions in the forthcoming election for 'Politics Now'. How would you select the sample of people you are going to ask?

If you stood in the town centre in the middle of one morning and asked passers-by, you would probably get an unduly high proportion of those who, for one reason or another, were not employed. It is quite possible that this group has different voting intentions from those in work.

If you selected names from the telephone directory, you would automatically exclude those who do not have telephones, those who do not have landlines and those who are ex-directory.

It is actually very difficult to come up with a plan that will yield a fair sample, one that is not biased in some direction or another. There are, however, a number of established sampling techniques and these are described in the next section of this chapter.

?

- Each of the situations below involves a population and a sample. In each case identify both, briefly but precisely.
  - (i) A member of parliament is interested in whether her constituents support proposed legislation to make convicted drug dealers do hard physical work every day while they are in prison. Her staff report that letters on the proposed legislation have been received from 361 constituents of whom 309 support it.
  - (ii) A flour company wants to know what proportion of households in Karachi bake some or all of their own bread. A sample of 500 residential addresses in Karachi is taken and interviewers are sent to these addresses. The interviewers are employed during regular working hours on weekdays and interview only during these hours.
  - (iii) The Chicago Police Department wants to know how black residents of Chicago feel about police service. A questionnaire with several questions about the police is prepared. A sample of 300 postal addresses in predominantly black areas of Chicago is taken and a police officer is sent to each address to administer the questionnaire to an adult living there.
- Each sampling situation contains a serious source of probable bias. In each case give the reason that bias may occur and also the direction of the bias.

1.3 Sampling techniques

In considering the following techniques, it is worth repeating that a key aim when taking a sample is to obtain a sample that is **representative** of the parent population being investigated. It is assumed that the sampling is done without replacement, otherwise, for example, one person could give an opinion twice, or more. The fraction of the population that is selected is called the sampling fraction.

» Sampling fraction = $\dfrac{\text{sample size}}{\text{population size}}$

Random sampling

In a **random sampling procedure**, every member of the population may be selected; there is a non-zero probability of this happening (and, of course, the probability is less than 1). In many random sampling procedures, for example drawing names out of a hat, every member of the population has an equal probability of being selected.

In a **simple random sampling** procedure, every possible sample of a given size is equally likely to be selected. It follows that in such a procedure every member of the parent population is equally likely to be selected. However, the converse is not true. It is possible to devise a sampling procedure in which every member is equally likely to be selected but some samples are not permissible.

?

- A school has 20 classes, each with 30 students. One student is chosen at random from each class, giving a sample size of 20. Why is this not a simple random sampling procedure?
- If you write the name of each student in the school on a slip of paper, put all the slips in a box, shake it well and then take out 20, would this be a simple random sample?

Simple random sampling is fine when you can do it, but you must have a sampling frame. The selection of items within the frame is often done using tables of random numbers.

Using random numbers

Usually, each element in the frame is given a number, starting at 1. You then select elements for the sample using random number tables or the random number generator on a calculator or computer.

Suppose that you need to select a sample of 15 houses from a numbered list of 483 houses. Using random number tables, you choose a random starting position and take the digits in groups of three. If the first set of three digits is 247, you put house number 247 from the list into your sample. If the next number is 832, you ignore it because it does not correspond to a house in the list. You continue in this way until you have a sample of 15 houses. (If any number occurs more than once, you still only include it once in your sample.)

In some circumstances, you might choose to assign random numbers in a less wasteful way. For example, you could subtract 500 from any random numbers above 500, so instead of discarding 832 you would choose house (832 – 500) = 332. Whether this is worthwhile depends on the sample size and the method being used to link the numbers to the elements in the sampling frame.

When using a random number generator on a calculator, you use the same procedure. If the calculator only provides three digits and you need five, you can generate two sets of three digits and discard the last digit.

ACTIVITY 1.1

Using the random numbers below, which items would you choose from a numbered list of the 17 841 inhabitants of a town if you want a random sample of size 10? Start with the top-left random number and work along each row in order.

14592
12471

54	66	35	88	98	91	45	92	12	47
12	16	71	83	94	22	44	57	43	43
45	32	26	37	19	89	27	02	77	14
85	98	46	56	50	71	07	65	33	63
51	63	71	95	36	36	17	77	53	40
25	95	65	04	59	80	16	59	21	43
91	55	88	14	82	48	48	94	38	34
60	87	82	35	35	45	45	08	44	37

e

Other sampling techniques

There are many other sampling techniques. **Survey design**, the formulation of the most appropriate sampling procedure in a particular situation, is a major topic within statistics.

Stratified sampling

You have already thought about the difficulty of conducting a survey of people's voting intentions in a particular area before an election. In that situation it is possible to identify a number of different sub-groups that you might expect to have different voting patterns: low-, medium- and high-income groups; urban, suburban and rural dwellers; young, middle-aged and elderly voters; men and women; and so on. The sub-groups are called **strata**.

In **stratified sampling**, you would ensure that all strata were sampled. You would need to sample from high income, suburban, elderly women; medium income, rural young men; etc. In this example, 54 strata ($3 \times 3 \times 3 \times 2$) have been identified. If the numbers sampled in the various strata are proportional to the size of their populations, the procedure is called **proportional stratified sampling**. If the sampling is not proportional, then appropriate weighting has to be used.

The selection of the items to be sampled within each stratum is usually done by simple random sampling. Stratified sampling will usually lead to more accurate results about the entire population, and will also give useful information about the individual strata.

Cluster sampling

Cluster sampling also starts with sub-groups, or strata, of the population, but in this case the items are chosen from one or several of the sub-groups. The sub-groups are now called clusters. It is important that each cluster should be reasonably representative of the entire population. If, for example, you were asked to investigate the incidence of a particular parasite in the puffin population of Northern Europe, it would be impossible to use simple random sampling. Rather you would select a number of sites and then catch some puffins at each place. This is cluster sampling. Instead of selecting from the whole population you are choosing from a limited number of clusters.

Systematic sampling

Systematic sampling is a method of choosing individuals from a sampling frame. If you were surveying telephone subscribers, you might select a number at random, say 66, and then sample the 66th name on every page of the directory. If the items in the sampling frame are numbered 1, 2, 3, ..., you might choose a random starting point like 38 and then sample numbers 38, 138, 238, and so on.

When using systematic sampling you have to beware of any cyclic patterns within the frame. For example, suppose a school list is made up class by class, each of exactly 25 children, in order of merit, so that numbers 1, 26, 51, 76, 101, ..., in the frame are those at the top of their class. If you sample every 50th child starting with number 26, you will conclude that the children in the school are very bright.

Quota sampling

Quota sampling is the method often used by companies employing people to carry out opinion surveys. An interviewer's quota is always specified in stratified terms: how many males and how many females, etc. The choice of who is sampled is then left up to the interviewer and so is definitely non-random.

Opportunity sampling

As its name suggests, **opportunity sampling** is used when circumstances make a sample readily available. As an example, the delegates at a conference of hospital doctors are used as a sample to investigate the opinion of hospital doctors in general on a particular issue. This can obviously bias the results and opportunity sampling is often viewed as the weakest form of sampling; however, it can be useful for social scientists who want to study behaviours of particular groups of people, such as criminals, where research will lead to individual case studies rather than results that are applied to the whole population. Opportunity sampling can also be useful for an initial pilot study before a wider investigation is carried out.

Self-selecting sample

A sample is **self-selected** when those involved volunteer to take part, or are given the choice to participate or decline. Examples are an online survey or when a researcher on the street asks people if they want to take part in a survey; they can agree to answer questions or say no. If enough people say no then this may affect the results of the survey. If someone who agrees to take part in the survey then refuses to answer some of the questions, this is also a form of self-selection that can bias the results.

It is also possible to have a self-selecting sample when people volunteer to answer questions, such as a text-in survey on a radio station to identify the 'best' song; however, as this can be easily identified as a self-selecting survey, it is not usually used in serious research.

Exercise 1A

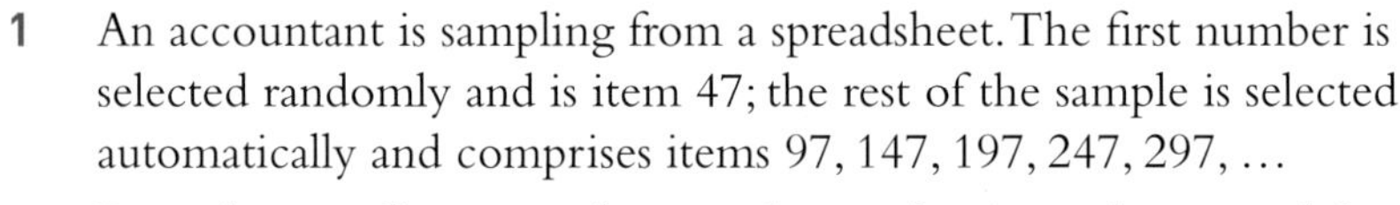

1 An accountant is sampling from a spreadsheet. The first number is selected randomly and is item 47; the rest of the sample is selected automatically and comprises items 97, 147, 197, 247, 297, …

Does the sampling procedure produce a simple random sample?

2 Pritam is a student at Avonford High School. He is given a list of all the students in the school. The list numbers the students from 1 to 2500. Pritam generates a four-digit random number between 0 and 1 on his calculator, for example 0.4325. He multiplies the random number by 2500 and notes the integer part. For example, 0.4325×2500 results in 1081 so Pritam chooses the student listed as 1081. He repeats the process until he has a sample of 100 names.

(i) What type of sampling procedure is Pritam carrying out?

(ii) What is the sampling fraction in this case?

3 Mr Jones wishes to find out if a mobile grocery service would be popular in Avonford. Describe how Mr Jones can generate a random sample of addresses to call at to seek the residents' views.

4 Teegan is trying to encourage people to shop at her boutique. She has produced a short questionnaire and asks her first 50 customers one day about their fashion preferences. Criticise Teegan's sampling method.

5 Marie wants to choose one student at random from Anthea, Bill and Charlie. She throws two fair coins. If both coins show tails she will choose Anthea. If both coins show heads she will choose Bill. If the coins show one of each she will choose Charlie.

(i) Explain why this is not a fair method for choosing the student.

(ii) Describe how Marie could use the two coins to give a fair method for choosing the student.

Cambridge International AS & A Level Mathematics
9709 Paper 71 Q1 June 2013

e

CP

6 Identify the sampling procedures that would be appropriate in the following situations.

(i) A local education officer wishes to estimate the mean number of children per family on a large housing estate.

(ii) A consumer protection body wishes to estimate the proportion of trains that are running late.

(iii) A marketing consultant wishes to investigate the proportion of households in a town that have a personal computer.

(iv) A local politician wishes to carry out a survey into people's views on capital punishment within your area.

(v) A health inspector wishes to investigate what proportion of people wear spectacles.

(vi) Ministry officials wish to estimate the proportion of cars with bald tyres.

(vii) A television company wishes to estimate the proportion of householders who have not paid their television licence fee.

(viii) The police want to find out how fast cars travel in the outside lane of a motorway.

(ix) A sociologist wants to know how many girlfriends the average 18-year-old boy has had.

(x) The headteacher of a large school wishes to estimate the average number of hours of homework done per week by the students.

KEY POINTS

1 There are essentially two reasons why you might wish to take a sample:
   - to estimate the values of the parameters of the parent population
   - to conduct a hypothesis test.
2 When taking a sample you should ensure that:
   - the data are relevant
   - the data are unbiased
   - the data are not distorted by the act of collection
   - a suitable person is collecting the data
   - the sample is of a suitable size
   - a suitable sampling procedure is being followed.
3 In a random sample, every member of the population has a non-zero probability of being selected. In many random sampling procedures, every member of the population has an equal probability of being selected.
4 In simple random sampling, every possible sample of a given size has an equal probability of being selected.

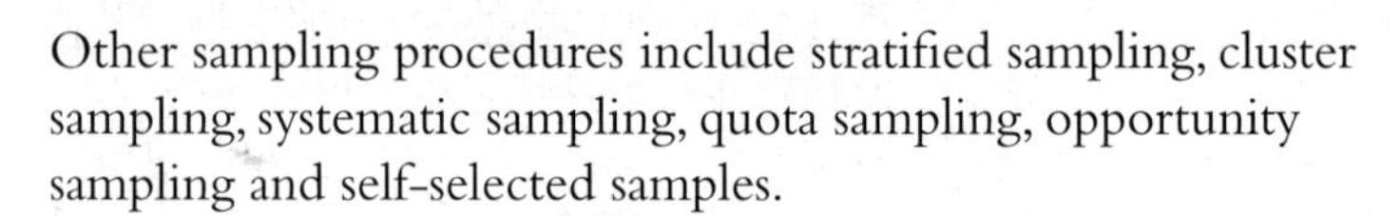

Other sampling procedures include stratified sampling, cluster sampling, systematic sampling, quota sampling, opportunity sampling and self-selected samples.

LEARNING OUTCOMES

Now that you have finished this chapter, you should be able to
- understand the terms:
  - population
  - sample
  - random sample
- understand the necessity for a sample to be random
- criticise a given sampling method.

2 Continuous random variables

A theory is a good theory if it satisfies two requirements: It must accurately describe a large class of observations on the basis of a model that contains only a few arbitrary elements, and it must make definite predictions about the results of future observations.
Stephen Hawking (1942–)
A Brief History of Time

Lucky escape for local fisherman

Local fisherman Zhang Wei stared death in the face yesterday as he was plucked from his boat by a freak wave. Only the quick thinking of his brother Xiuying who grabbed hold of his legs, saved Wei from a watery grave.

'It was a bad day and suddenly this lump of water came down on us,' said Wei. 'It was a wave in a million. It must have been higher than our house, which is about 11 m high, and it caught me off guard'.

Hero Xiuying is a man of few words. 'All in the day's work' was his only comment.

Freak waves do occur and they can have serious consequences in terms of damage to shipping, oil rigs and coastal defences, sometimes resulting in loss of life. It is important to estimate how often they will occur, and how high they will be.

?

› Was Zhang Wei's one in a million estimate for a wave higher than 11 metres at all accurate?

Before you can answer this question, you need to know the **probability density** of the heights of waves at that time of the year in the area where the Zhang brothers were fishing. The graph in Figure 2.1 shows this sort of information; it was collected in the same season of the year as the Zhang accident.

To obtain Figure 2.1 a very large amount of wave data had to be collected. This allowed the class interval widths of the wave heights to be sufficiently small for the outline of the curve to acquire this shape. It also ensured that the sample data were truly representative of the population of waves at that time of the year.

In a graph such as Figure 2.1 the vertical scale is a measure of probability density. Probabilities are found by estimating the area under the curve. The total area is 1.0, meaning that effectively all waves at this place have heights between 0.6 and 12.0 m (see Figure 2.2).

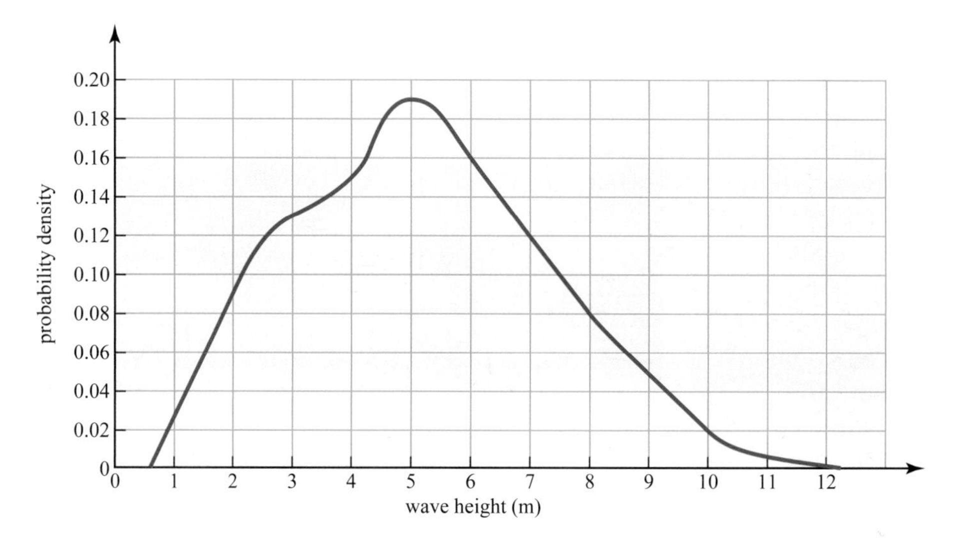

▲ **Figure 2.1**

If this had been the place where the Zhang brothers were fishing, the probability of encountering a wave at least 11 m high would have been 0.003, about 1 in 300 (see Figure 2.2). Clearly Wei's description of it as 'a wave in a million' was not justified purely by its height. The fact that he called it a 'lump of water' suggests that perhaps it may have been more remarkable for its steep sides than its height.

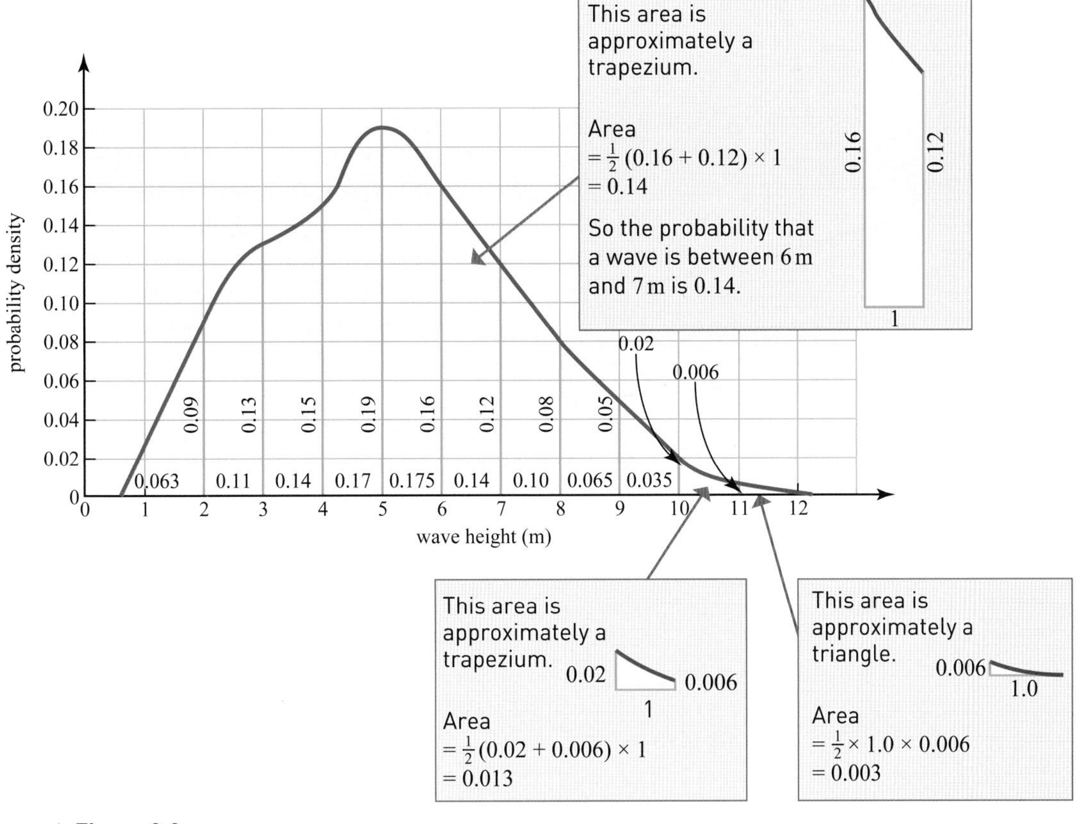

▲ **Figure 2.2**

2.1 Probability density function

In the wave height example the curve was determined experimentally. The curve is continuous because the random variable, the wave height, is continuous and not discrete. The possible heights of waves are not restricted to particular steps (say, every $\frac{1}{2}$ metre), but may take any value within a range.

> Is it reasonable to describe the height of a wave as **random**?

A function represented by a curve of this type is called a **probability density function**, often abbreviated to PDF The probability density function of a continuous random variable, X, is usually denoted by $f(x)$. If $f(x)$ is a PDF it follows that:

- $f(x) \geqslant 0$ for all x — You cannot have negative probabilities.
- $\int_{\text{All values of } x} f(x)\,dx = 1$ — The total area under the curve is 1.

For a continuous random variable with probability density function $f(x)$, the probability that X lies in the interval $[a, b]$ is given by

$$P(a \leqslant X \leqslant b) = \int_a^b f(x)\,dx$$

Looking at Figure 2.1, you will see that in this case, the probability density function has quite a complicated curve and so it is not possible to find a simple algebraic expression with which to model it.

Most of the techniques in this chapter assume that you do in fact have a convenient algebraic expression with which to work. However, the methods are still valid if this is not the case, but you would need to use numerical, rather than algebraic, techniques for integration and differentiation. In the high-wave incident mentioned on pages 11–12, the areas corresponding to wave heights of less than 2 m and of at least 11 m were estimated by treating the shape as a triangle: other areas were approximated by trapezia.

Note: Class boundaries

If you were to ask the question 'What is the probability of a randomly selected wave being exactly 2 m high?' the answer would be zero. If you measured a likely looking wave to enough decimal places (assuming you could do so), you would eventually come to a figure that was not zero. The wave height might be 2.01... m or 2.000 003...m but the probability of it being exactly 2 m is infinitesimally small. Consequently, in theory it makes no difference whether you describe the class interval from 2 to 2.5 m as $2 < h < 2.5$ or as $2 \leqslant h \leqslant 2.5$.

However, in practice, measurements are always rounded to some extent. The reality of measuring a wave's height means that you would probably be quite happy to record it to the nearest 0.1 m and get on with the next wave. So, in practice, measurements of 2.0 m and 2.5 m probably will be recorded, and intervals have to be defined so that it is clear which class they belong to. You would normally expect $<$ at one end of the interval and $\leqslant$ at the other: either $2 \leqslant h < 2.5$ or $2 < h \leqslant 2.5$. In either case the probability of the wave being within the interval would be given by

$$\int_{2}^{2.5} \mathrm{f}(x)\,\mathrm{d}x$$

Somewhere an empty-pocketed thief is nursing a sore leg and regretting the loss of a pair of trousers. Security guard Matthias Arrigo, and Fido, a Jack Russell, were doing a late-night check round the office when they came upon the intruder on the ground floor.

'I didn't need to say anything,' Matthias told me; 'Fido went straight for him and grabbed him by the leg.' After a tussle the man's trousers tore, leaving Fido with a mouthful of material while the man made good his escape out of the window.

Following the incident, the office management is looking at an electronic security system. 'Fido won't live for ever,' explained Matthias.

Example 2.1

Matthias is thinking of fitting an electronic security system inside the offices. He has been told by a manufacturer that the lifetime, X years, of the system he has in mind has the PDF.

$$f(x) = \frac{3x(20-x)}{4000} \quad \text{for } 0 \leqslant x \leqslant 20,$$

and $\quad f(x) = 0 \quad$ otherwise.

(i) Show that the manufacturer's statement is consistent with f(x) being a probability density function.

(ii) Find the probability that

(a) it fails in the first year

(a) it lasts 10 years but then fails in the next year.

Solution

(i) The condition $f(x) \geqslant 0$ for all values of x between 0 and 20 is satisfied, as shown by the graph of f(x), Figure 2.3.

exact P(x) = 0

2 & 3

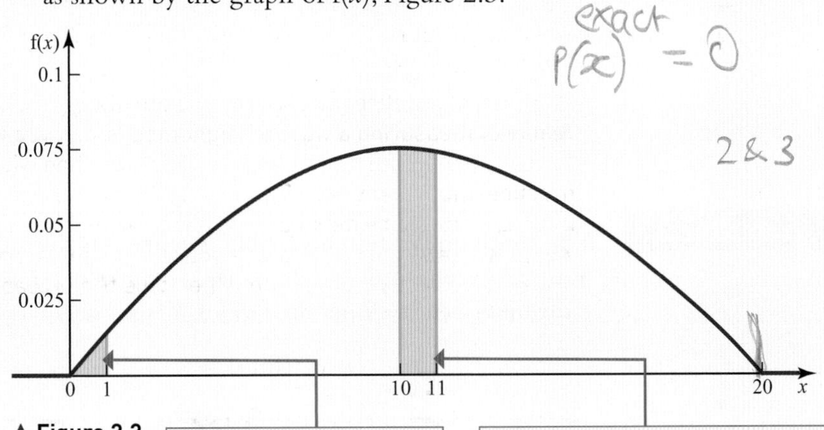

▲ **Figure 2.3**

The other condition is that the area under the curve is 1.

$$\text{Area} = \int_{-\infty}^{\infty} f(x)\,dx = \int_0^{20} \frac{3x(20-x)}{4000}\,dx$$

$$= \frac{3}{4000}\int_0^{20}(20x - x^2)\,dx$$

$$= \frac{3}{4000}\left[10x^2 - \frac{x^3}{3}\right]_0^{20}$$

$$= \frac{3}{4000}\left[10 \times 20^2 - \frac{20^3}{3}\right]$$

$= 1$, as required. →

(ii) (a) *It fails in the first year.*

This is given by $P(X < 1) = \int_0^1 \frac{3x(20-x)}{4000}\,dx$

$$= \frac{3}{4000}\int_0^1 (20x - x^2)\,dx$$

$$= \frac{3}{4000}\left[10x^2 - \frac{x^3}{3}\right]_0^1$$

$$= \frac{3}{4000}\left(10\times 1^2 - \frac{1^3}{3}\right)$$

$$= 0.007\,25$$

(b) *It fails in the 11th year.*

This is given by $P(10 \leqslant X < 11)$

$$= \int_{10}^{11} \frac{3x(20-x)}{4000}\,dx$$

$$= \frac{3}{4000}\left[10x^2 - \tfrac{1}{3}x^3\right]_{10}^{11}$$

$$= \frac{3}{4000}\left(10\times 11^2 - \tfrac{1}{3}\times 11^3\right) - \frac{3}{4000}\left(10\times 10^2 - \tfrac{1}{3}\times 10^3\right)$$

$$= 0.074\,75$$

Example 2.2

The continuous random variable X represents the amount of sunshine in hours between noon and 4 pm at a skiing resort in the high season. The probability density function, $f(x)$, of X is modelled by

$$f(x) = \begin{cases} kx^2 & \text{for } 0 \leqslant x \leqslant 4 \\ 0 & \text{otherwise.} \end{cases}$$

(i) Find the value of k.

(ii) Find the probability that on a particular day in the high season there is more than two hours of sunshine between noon and 4 pm.

Solution

(i) To find the value of k you must use the fact that the area under the graph of $f(x)$ is equal to 1.

$$\int_{-\infty}^{\infty} f(x)\,dx = \int_0^4 kx^2\,dx = 1$$

Therefore $\left[\frac{kx^3}{3}\right]_0^4 = 1$

$$\frac{64k}{3} = 1$$

So $k = \frac{3}{64}$

(ii)

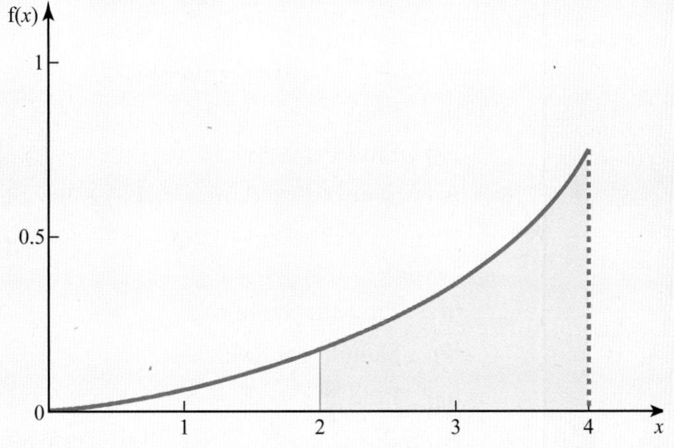

▲ **Figure 2.4**

The probability of more than 2 hours of sunshine is given by

$$\begin{aligned}
P(X > 2) &= \int_2^{\infty} f(x)\,dx = \int_2^4 \frac{3x^2}{64}\,dx \\
&= \left[\frac{x^3}{64}\right]_2^4 \\
&= \frac{64-8}{64} \\
&= \frac{56}{64} \\
&= 0.875
\end{aligned}$$

Exercise 2A

1 The continuous random variable X has probability density function f(x) where

$$\begin{aligned} f(x) &= kx \quad &\text{for } 1 \leqslant x \leqslant 6 \\ &= 0 &\text{otherwise.} \end{aligned}$$

(i) Find the value of the constant k.

(ii) Sketch $y = f(x)$.

(iii) Find $P(X > 5)$.

(iv) Find $P(2 \leqslant X \leqslant 3)$.

2 The continuous random variable X has PDF f(x) where

$$\begin{aligned} f(x) &= k(5 - x) \quad &\text{for } 0 \leqslant x \leqslant 4 \\ &= 0 &\text{otherwise.} \end{aligned}$$

(i) Find the value of the constant k.

(ii) Sketch $y = f(x)$.

(iii) Find $P(1.5 \leqslant X \leqslant 2.3)$.

3 The continuous random variable X has PDF f(x) where

$$\begin{aligned} f(x) &= ax^3 \quad &\text{for } 0 \leqslant x \leqslant 3 \\ &= 0 &\text{otherwise.} \end{aligned}$$

(i) Find the value of the constant a.

(ii) Sketch $y = f(x)$.

(iii) Find $P(X \leqslant 2)$.

4 The continuous random variable X has PDF $f(x)$ where

$$f(x) = c \quad \text{for } -3 \leqslant x \leqslant 5$$
$$= 0 \quad \text{otherwise.}$$

(i) Find c.

(ii) Sketch $y = f(x)$.

(iii) Find $P(|X| < 1)$.

(iv) Find $P(|X| > 2.5)$

5 A continuous random variable X has PDF $f(x)$ where

$$f(x) = k(x-1)(6-x) \quad \text{for } 1 \leqslant x \leqslant 6$$
$$= 0 \quad \text{otherwise.}$$

(i) Find the value of k.

(ii) Sketch $y = f(x)$.

(iii) Find $P(2 \leqslant X \leqslant 3)$.

M **6** A random variable X has PDF $f(x)$ where

$$f(x) = kx(3-x) \quad \text{for } 0 \leqslant x \leqslant 3$$
$$= 0 \quad \text{otherwise.}$$

(i) Find the value of k.

(ii) The lifetime (in years) of an electronic component is modelled by this distribution. Two such components are fitted in a radio that will only function if both devices are working. Find the probability that the radio will still function after two years, assuming that their failures are independent.

M **7** The planning officer in a city hall needs information about how long cars stay in the car park, and asks the attendant to do a check on the times of arrival and departure of 100 cars. The attendant provides the following data.

Length of stay	Under 1 hour	1–2 hours	2–4 hours	4–10 hours	More than 10 hours
Number of cars	20	14	32	34	0

The planning officer suggests that the length of stay in hours may be modelled by the continuous random variable X with PDF of the form

$$f(x) = k(20-2x) \quad \text{for } 0 \leqslant x \leqslant 10$$
$$= 0 \quad \text{otherwise.}$$

(i) Find the value of k.

(ii) Sketch the graph of $f(x)$.

(iii) According to this model, how many of the 100 cars would be expected to fall into each of the four categories?

(iv) Do you think the model fits the data well?

(v) Are there any obvious weaknesses in the model? If you were the planning officer, would you be prepared to accept the model as it is, or would you want any further information?

M **8** A fish farmer has a very large number of trout in a lake. Before deciding whether to net the lake and sell the fish, she collects a sample of 100 fish and weighs them. The results (in kg) are as follows.

Weight, W	**Frequency**
$0 < W \leqslant 0.5$	2
$0.5 < W \leqslant 1.0$	10
$1.0 < W \leqslant 1.5$	23
$1.5 < W \leqslant 2.0$	26

Weight, W	**Frequency**
$2.0 < W \leqslant 2.5$	27
$2.5 < W \leqslant 3.0$	12
$3.0 < W$	0

(i) Illustrate these data on a histogram, with the number of fish on the vertical scale and W on the horizontal scale. Is the distribution of the data symmetrical, positively skewed or negatively skewed?

A friend of the farmer suggests that W can be modelled as a continuous random variable and proposes four possible probability density functions.

$$\mathrm{f}_1(w) = \tfrac{2}{9}w(3-w) \qquad \mathrm{f}_2(w) = \tfrac{10}{81}w^2(3-w)^2$$

$$\mathrm{f}_3(w) = \tfrac{4}{27}w^2(3-w) \qquad \mathrm{f}_4(w) = \tfrac{4}{27}w(3-w)^2$$

in each case for $0 < W \leqslant 3$.

(ii) Sketch the curves of the four PDFs and state which one matches the data most closely in general shape.

(iii) Use this PDF to calculate the number of fish which that model predicts should fall within each group.

(iv) Do you think it is a good model?

CP **9** A random variable X has a probability density function $\mathrm{f}(x)$ given by

$$\begin{aligned} \mathrm{f}(x) &= cx(8-x) \qquad && 0 \leqslant x \leqslant 8 \\ &= 0 && \text{otherwise.} \end{aligned}$$

(i) Show that $c = \frac{3}{256}$.

(ii) The lifetime X (in years) of an electric light bulb has this distribution. Given that a standard lamp is fitted with two such new bulbs and that their failures are independent, find the probability that neither bulb fails in the first year and the probability that exactly one bulb fails within two years.

$c\int x(8-x)$

$8x - 8x$

$c\left(4x^2 - \frac{x^3}{3}\right)$

sub in 8

[illegible] $c\,\frac{256}{3} = 1$

$c = 3/256$

M

10 This graph shows the probability density function, $f(x)$, for the heights, X, of waves at the point with Latitude 44°N Longitude 41°W.

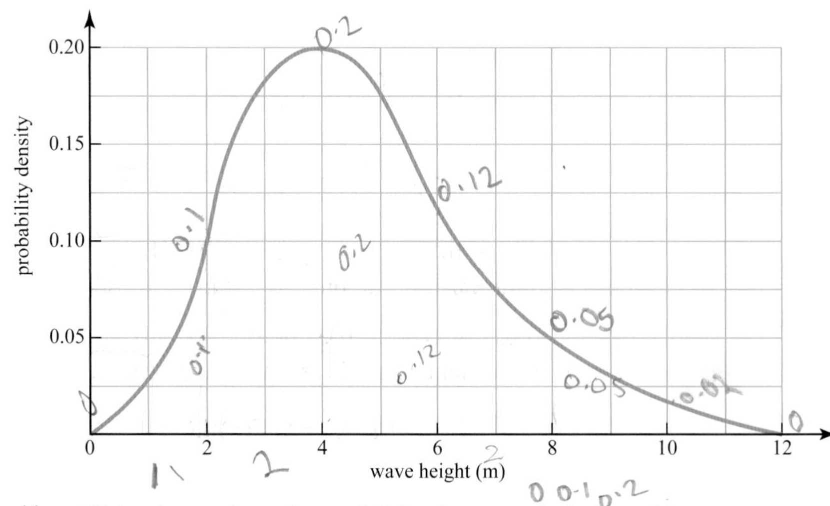

(i) Write down the values of $f(x)$ when $x = 0, 2, 4, \ldots, 12$.

(ii) Hence estimate the probability that the height of a randomly selected wave is in the interval

(a) 0–2 m (b) 2–4 m (c) 4–6 m
(d) 6–8 m (e) 8–10 m (f) 10–12 m.

A model is proposed in which

$$f(x) = kx(12 - x)^2 \quad \text{for } 0 \leqslant x \leqslant 12$$
$$= 0 \quad \text{otherwise.}$$

(iii) Find the value of k.

(iv) Find, according to this model, the probability that a randomly selected wave is in the interval

(a) 0–2 m (b) 2–4 m (c) 4–6 m
(d) 6–8 m (e) 8–10 m (f) 10–12 m.

(v) By comparing the figures from the model with the real data, state whether or not you think it is a good model.

PS

11 A continuous random variable X has PDF $f(x)$ where

$$f(x) = \frac{k}{x^3} \quad \text{for } x \geqslant 1$$
$$= 0 \quad \text{otherwise.}$$

(i) Find the value of k.

(ii) Sketch the graph of $y = f(x)$.

(iii) Find the value of m such that $P(X \leqslant m) = 0.75$.

PS

12 A continuous random variable X has PDF $f(x)$ where

$$f(x) = \frac{4k}{x^2} \quad \text{for } x \geqslant 1$$
$$= 0 \quad \text{otherwise.}$$

(i) Find the value of k.

(ii) The lifetime (X in years) of an electrical component has this distribution. Find the time after which 90% of the components will have failed.

2.2 Mean and variance

You will recall from *Probability & Statistics 1* that, for a discrete random variable, **mean** and **variance** are given by

$$\mu = \mathrm{E}(X) = \sum_i x_i p_i$$

$$\mathrm{Var}(X) = \sum_i (x_i - \mu)^2 p_i = \sum_i x_i^2 p_i - [\mathrm{E}(X)]^2$$

where μ is the mean and p_i is the probability of the outcome x_i for $i = 1, 2, 3, \ldots$, with the various outcomes covering all possibilities.

The expressions for the mean and variance of a continuous random variable are equivalent, but with summation replaced by integration.

$$\mu = \mathrm{E}(X) = \int_{\substack{\text{All}\\\text{values}\\\text{of } x}} x\,\mathrm{f}(x)\,\mathrm{d}x$$

$$\mathrm{Var}(X) = \int_{\substack{\text{All}\\\text{values}\\\text{of } x}} (x - \mu)^2 \mathrm{f}(x)\,\mathrm{d}x = \int_{\substack{\text{All}\\\text{values}\\\text{of } x}} x^2 \mathrm{f}(x)\,\mathrm{d}x - [\mathrm{E}(X)]^2$$

$\mathrm{E}(X)$ is the same as the population mean, μ, and is often called the mean of X.

Example 2.3

The response time, in seconds, for a contestant in a general knowledge quiz is modelled by a continuous random variable X whose PDF is

$$\mathrm{f}(x) = \frac{x}{50} \quad \text{for } 0 < x \leqslant 10.$$

The rules state that a contestant who makes no answer is disqualified from the whole competition. This has the consequence that everybody gives an answer, if only a guess, to every question. Find

(i) the mean time in seconds for a contestant to respond to a particular question

(ii) the standard deviation of the time taken.

The organiser estimates the proportion of contestants who are guessing by assuming that they are those whose time is at least one standard deviation greater than the mean.

(iii) Using this assumption, estimate the probability that a randomly selected response is a guess. ➜

Solution

(i) Mean time: $\mathrm{E}(X) = \int_0^{10} x\,\mathrm{f}(x)\,\mathrm{d}x$

$$= \int_0^{10} \frac{x^2}{50}\,\mathrm{d}x$$

$$= \left[\frac{x^3}{150}\right]_0^{10} = \frac{1000}{150} = \frac{20}{3}$$

$$= 6\tfrac{2}{3}$$

The mean time is $6\frac{2}{3}$ seconds.

(ii) Variance: $\mathrm{Var}(X) = \int_0^{10} x^2\,\mathrm{f}(x)\,\mathrm{d}x - [\mathrm{E}(X)]^2$

$$= \int_0^{10} \frac{x^3}{50}\,\mathrm{d}x - \left(6\tfrac{2}{3}\right)^2$$

$$= \left[\frac{x^4}{200}\right]_0^{10} - \left(6\tfrac{2}{3}\right)^2$$

$$= 5\tfrac{5}{9}$$

Standard deviation $= \sqrt{\text{variance}} = \sqrt{5.\dot{5}}$

The standard deviation of the times is 2.357 seconds (to 3 d.p.).

(iii) All those with response times greater than $6.667 + 2.357 = 9.024$ seconds are taken to be guessing. The longest possible time is 10 seconds.

The probability that a randomly selected response is a guess is given by

$$\int_{9.024}^{10} \frac{x}{50}\,\mathrm{d}x = \left[\frac{x^2}{100}\right]_{9.024}^{10}$$

$$= 0.186$$

So just under 1 in 5 answers are deemed to be guesses.

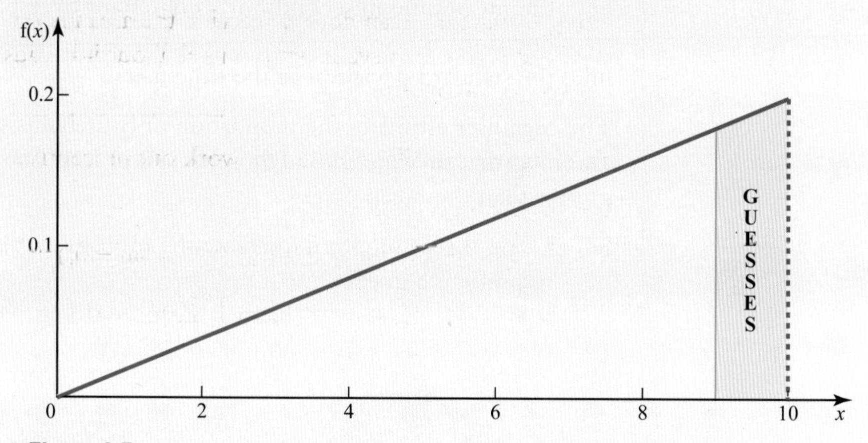

▲ **Figure 2.5**

Note

Although the intermediate answers have been given rounded to three decimal places, more figures have been carried forward into subsequent calculations.

2.3 The median

The median value of a continuous random variable X with PDF $\mathrm{f}(x)$ is the value m for which

$$P(X < m) = P(X > m) = 0.5.$$

Consequently $\int_{-\infty}^{m} \mathrm{f}(x)\,\mathrm{d}x = 0.5$ and $\int_{m}^{\infty} \mathrm{f}(x)\,\mathrm{d}x = 0.5.$

The median is the value m such that the line $x = m$ divides the area under the curve $\mathrm{f}(x)$ into two equal parts. In Figure 2.6 a is the smallest possible value of X, b the largest. The line $x = m$ divides the shaded region into two regions A and B, both with area 0.5.

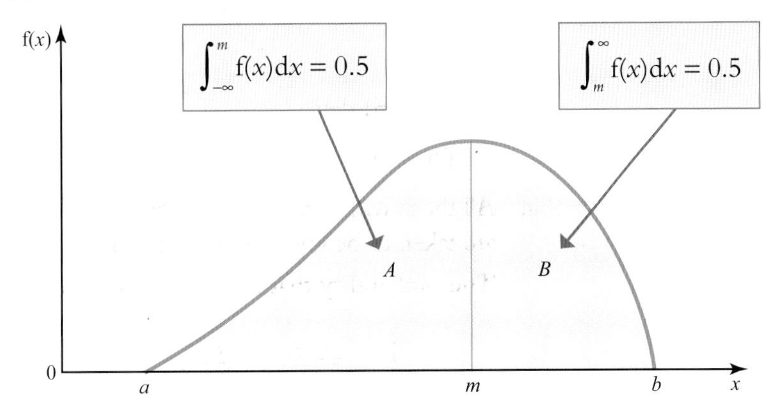

▲ **Figure 2.6**

In general, the mean does not divide the area into two equal parts but it will do so if the curve is symmetrical about it because, in that case, it is equal to the median.

You can use a similar method to work out percentiles.

For the 90th percentile

$$P(X < m) = 0.9 \text{ and } P(X > m) = 0.1$$

So $\int_{-\infty}^{m} \mathrm{f}(x)\,\mathrm{d}x = 0.9$ and $\int_{m}^{\infty} \mathrm{f}(x)\,\mathrm{d}x = 0.1.$

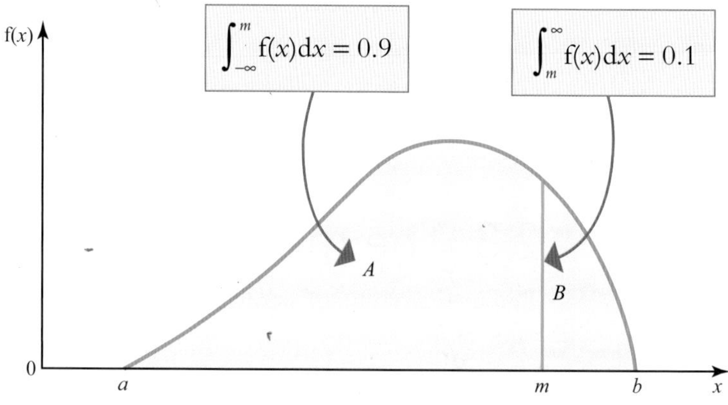

▲ **Figure 2.7**

2.4 The mode

The mode of a continuous random variable X whose PDF is $f(x)$ is the value of x for which $f(x)$ has the greatest value. Thus the mode is the value of X where the curve is at its highest.

If the mode is at a local maximum of $f(x)$, then it may often be found by differentiating $f(x)$ and solving the equation

$$f'(x) = 0.$$

› For which of the distributions in Figure 2.8 could the mode be found by differentiating the PDF?

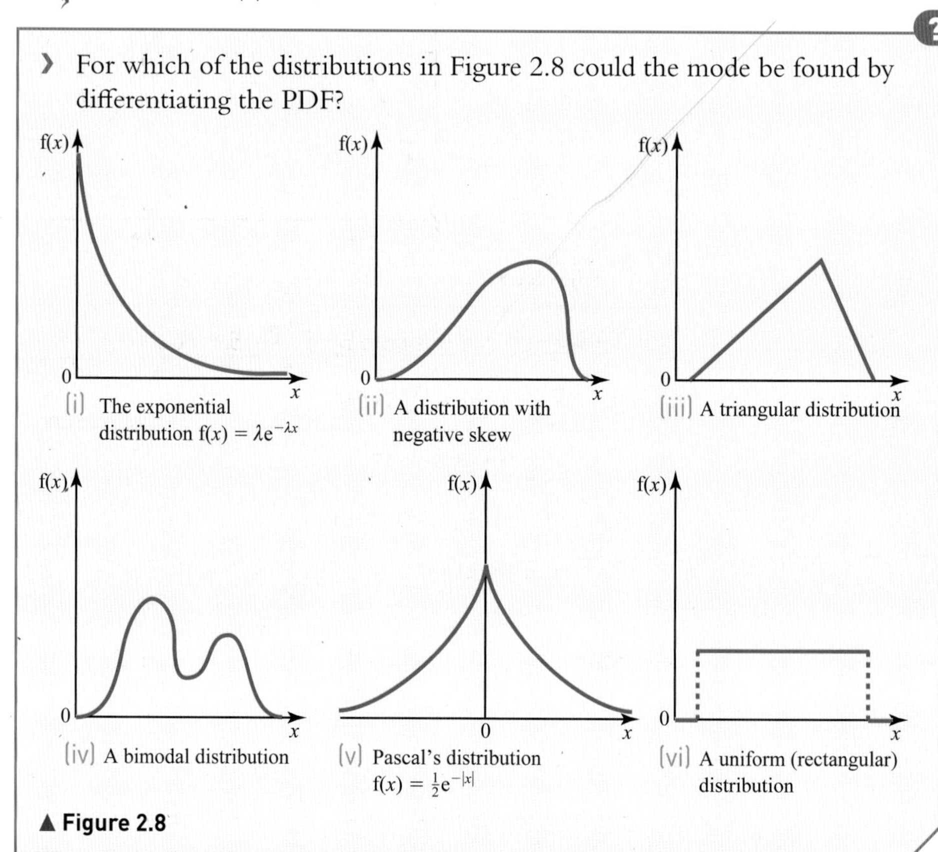

▲ **Figure 2.8**

Example 2.4

The continuous random variable X has PDF $f(x)$ where

$$f(x) = 4x(1 - x^2) \quad \text{for } 0 \leqslant x \leqslant 1$$
$$= 0 \quad \text{otherwise.}$$

Find

(i) the mode

(ii) the median.

Solution

$x = -0.577$ is also a root of $f'(x) = 0$ but is outside the range $0 \leqslant x \leqslant 1$.

(i) The mode is found by differentiating $f(x) = 4x - 4x^3$

$$f'(x) = 4 - 12x^2$$

Solving $f'(x) = 0$ $\quad x = \frac{1}{\sqrt{3}} = 0.577$ to 3 d.p.

It is easy to see from the shape of the graph (see Figure 2.9) that this must be a maximum, and so the mode is 0.577.

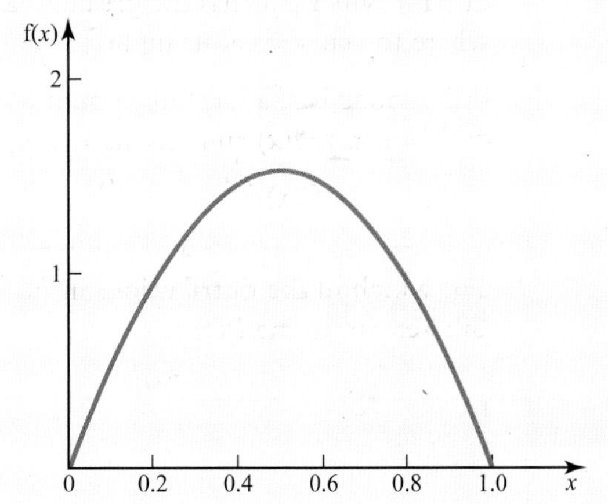

▲ **Figure 2.9**

(ii) The median, m, is given by $\int_{-\infty}^{m} f(x)\,dx = 0.5$

$$\Rightarrow \quad \int_{0}^{m} (4x - 4x^3)\,dx = 0.5$$

Since $x \geqslant 0$

$$[2x^2 - x^4]_0^m = 0.5$$

$$2m^2 - m^4 = 0.5$$

Rearranging gives

$$2m^4 - 4m^2 + 1 = 0.$$

This is a quadratic equation in m^2. The formula gives

$$m^2 = \frac{4 \pm \sqrt{16 - 8}}{4}$$

$$m = 0.541 \text{ or } 1.307 \text{ to 3 d.p.}$$

Since 1.307 is outside the domain of X, the median is 0.541.

2.5 The uniform (rectangular) distribution

It is common to describe distributions by the shapes of the graphs of their PDFs: U-shaped, J-shaped, etc.

The **uniform (rectangular) distribution** is particularly simple since its PDF is constant over a range of values and zero elsewhere.

In Figure 2.10, X may take values between a and b, and is zero elsewhere. Since the area under the graph must be 1, the height is $\frac{1}{b-a}$. The term 'uniform distribution' can be applied to both discrete and continuous variables so in the continuous case it is often written as 'uniform (rectangular)'.

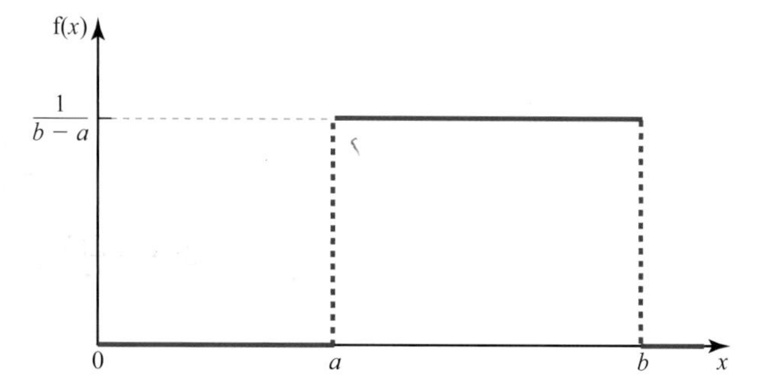

▲ **Figure 2.10**

Example 2.5

A junior gymnastics league is open to children who are at least five years old but have not yet had their ninth birthday. The age, X years, of a member is modelled by the uniform (rectangular) distribution over the range of possible values between five and nine. Age is measured in years and decimal parts of a year, rather than just completed years. Find

(i) the PDF f(x) of X

(ii) $P(6 \leqslant X \leqslant 7)$

(iii) $E(X)$

(iv) $Var(X)$

(v) the percentage of the children whose ages are within one standard deviation of the mean.

Solution

(i) The PDF $f(x) = \frac{1}{9-5} = \frac{1}{4}$ for $5 \leqslant x < 9$

$= 0$ otherwise.

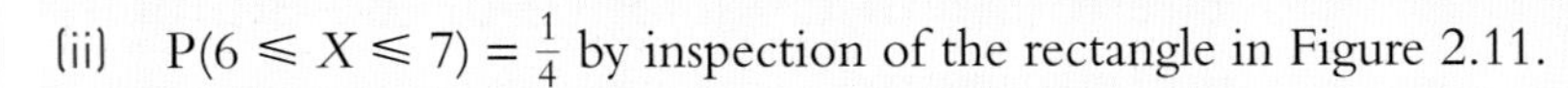

(ii) $P(6 \leqslant X \leqslant 7) = \frac{1}{4}$ by inspection of the rectangle in Figure 2.11.

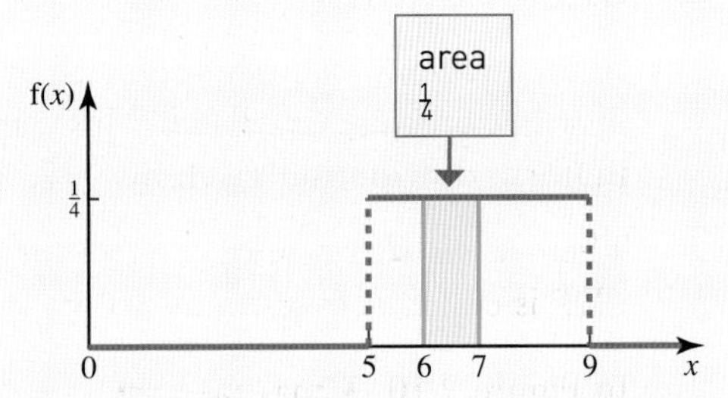

▲ **Figure 2.11**

Alternatively, using integration

$$P(6 \leqslant X \leqslant 7) = \int_6^7 f(x)\,dx = \int_6^7 \frac{1}{4}\,dx$$
$$= \left[\frac{x}{4}\right]_6^7$$
$$= \frac{7}{4} - \frac{6}{4}$$
$$= \frac{1}{4}.$$

(iii) By the symmetry of the graph, $E(X) = 7$. Alternatively, using integration

$$E(X) = \int_{-\infty}^{\infty} x f(x)\,dx = \int_5^9 \frac{x}{4}\,dx$$
$$= \left[\frac{x^2}{8}\right]_5^9$$
$$= \frac{81}{8} - \frac{25}{8} = 7.$$

(iv) $$\text{Var}(X) = \int_{-\infty}^{\infty} x^2 f(x)\,dx - [E(X)]^2 = \int_5^9 \frac{x^2}{4}\,dx - 7^2$$
$$= \left[\frac{x^3}{12}\right]_5^9 - 49$$
$$= \frac{729}{12} - \frac{125}{12} - 49$$
$$= 1.333 \text{ to 3 d.p.}$$

(v) Standard deviation $= \sqrt{\text{variance}} = \sqrt{1.333} = 1.155$.

So the percentage within 1 standard deviation of the mean is

$$\frac{2 \times 1.155}{4} \times 100\% = 57.7\%.$$

?

› What percentage would be within 1 standard deviation of the mean for a normal distribution? Why is the percentage less in this example?

The mean and variance of the uniform (rectangular) distribution

In the previous example the mean and variance of a particular uniform distribution were calculated. This can easily be extended to the general uniform distribution given by

$$\begin{aligned} f(x) &= \frac{1}{b-a} && \text{for } a \leqslant x \leqslant b \\ &= 0 && \text{otherwise.} \end{aligned}$$

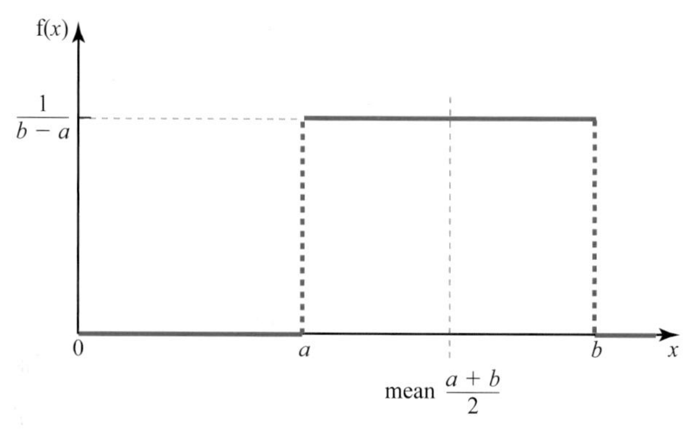

▲ **Figure 2.12**

Mean By symmetry the mean is $\frac{a+b}{2}$.

Variance

$$\begin{aligned} \text{Var}(X) &= \int_{-\infty}^{\infty} x^2 f(x)\,dx - [E(X)]^2 \\ &= \int_a^b x^2 f(x)\,dx - [E(X)]^2 \\ &= \int_a^b \frac{x^2}{b-a}dx - \left(\frac{a+b}{2}\right)^2 \\ &= \left[\frac{x^3}{3(b-a)}\right]_a^b - \frac{1}{4}(a^2+2ab+b^2) \\ &= \frac{b^3-a^3}{3(b-a)} - \frac{1}{4}(a^2+2ab+b^2) \\ &= \frac{(b-a)}{3(b-a)}(b^2+ab+a^2) - \frac{1}{4}(a^2+2ab+b^2) \\ &= \frac{1}{12}(b^2-2ab+a^2) \\ &= \frac{1}{12}(b-a)^2 \end{aligned}$$

Exercise 2B

1 The continuous random variable X has PDF $f(x)$ where

$$f(x) = \tfrac{1}{8}x \quad \text{for } 0 \leqslant x \leqslant 4$$
$$= 0 \quad \text{otherwise.}$$

Find

(i) $E(X)$

(ii) $Var(X)$

(iii) the median value of X.

2 The continuous random variable T has PDF defined by

$$f(t) = \frac{6-t}{18} \quad \text{for } 0 \leqslant t \leqslant 6$$
$$= 0 \quad \text{otherwise.}$$

Find

(i) $E(T)$

(ii) $Var(T)$

(iii) the median value of T.

3 The continuous random variable Y has PDF $f(y)$ defined by

$$f(y) = 12y^2(1-y) \quad \text{for } 0 \leqslant y \leqslant 1$$
$$= 0 \quad \text{otherwise.}$$

(i) Find $E(Y)$.

(ii) Find $Var(Y)$.

(iii) Verify that, to 2 decimal places, the median value of Y is 0.61.

4 The random variable X has PDF

$$f(x) = \tfrac{1}{6} \quad \text{for } -2 \leqslant x \leqslant 4$$
$$= 0 \quad \text{otherwise.}$$

(i) Sketch the graph of $f(x)$.

(ii) Find $P(X < 2)$.

(iii) Find $E(X)$.

(iv) Find $P(|X| < 1)$.

5 The continuous random variable X has PDF $f(x)$ defined by

$$f(x) = \begin{cases} \frac{2}{9}x(3-x) & \text{for } 0 \leqslant x \leqslant 3 \\ 0 & \text{otherwise.} \end{cases}$$

(i) Find $E(X)$.

(ii) Find $Var(X)$.

(iii) Find the mode of X.

(iv) Find the median value of X.

(v) Draw a sketch graph of $f(x)$ and comment on your answers to parts (i), (iii) and (iv) in the light of what it shows you.

6 The function $f(x) = \begin{cases} k(3+x) & \text{for } 0 \leqslant x \leqslant 2 \\ 0 & \text{otherwise} \end{cases}$

is the probability density function of the random variable X.

(i) Show that $k = \frac{1}{8}$.

(ii) Find the mean and variance of X.

(iii) Find the probability that a randomly selected value of X lies between 1 and 2.

7 A continuous random variable X has a uniform (rectangular) distribution over the interval $(4, 7)$. Find

(i) the PDF of X

(ii) $\mathrm{E}(X)$

(iii) $\mathrm{Var}(X)$

(iv) $\mathrm{P}(4.1 \leqslant X \leqslant 4.8)$.

M **8** The distribution of the lengths of adult Martian lizards is uniform between 10 cm and 20 cm. There are no adult lizards outside this range.

(i) Write down the PDF of the lengths of the lizards.

(ii) Find the mean and variance of the lengths of the lizards.

(iii) What proportion of the lizards have lengths within

(a) one standard deviation of the mean

(b) two standard deviations of the mean?

PS **9** The continuous random variable X has PDF $f(x)$ defined by

$$f(x) = \begin{cases} \dfrac{a}{x} & \text{for } 1 \leqslant x \leqslant 2 \\ 0 & \text{otherwise.} \end{cases}$$

(i) Find the value of a.

(ii) Sketch the graph of $f(x)$.

(iii) Find the mean and variance of X.

(iv) Find the proportion of values of X between 1.5 and 2.

(v) Find the median value of X.

10 The random variable X denotes the number of hours of cloud cover per day at a weather forecasting centre. The probability density function of X is given by

$$f(x) = \begin{cases} \dfrac{(x-18)^2}{k} & 0 \leqslant x \leqslant 24, \\ 0 & \text{otherwise,} \end{cases}$$

where k is a constant.

(i) Show that $k = 2016$.

(ii) On how many days in a year of 365 days can the centre expect to have less than 2 hours of cloud cover?

(iii) Find the mean number of hours of cloud cover per day.

Cambridge International AS & A Level Mathematics
9709 Paper 7 Q7 June 2005

11 The random variable X has probability density function given by

$$f(x) = \begin{cases} 4x^k & 0 \leqslant x \leqslant 1 \\ 0 & \text{otherwise,} \end{cases}$$

where k is a positive constant.

(i) Show that $k = 3$.

(ii) Show that the mean of X is 0.8 and find the variance of X.

(iii) Find the upper quartile of X.

(iv) Find the interquartile range of X.

Cambridge International AS & A Level Mathematics
9709 Paper 7 Q5 June 2006

12 If Usha is stung by a bee she always develops an allergic reaction. The time taken in minutes for Usha to develop the reaction can be modelled using the probability density function given by

$$f(t) = \begin{cases} \dfrac{k}{t+1} & 0 \leqslant t \leqslant 4, \\ 0 & \text{otherwise,} \end{cases}$$

where k is a constant.

(i) Show that $k = \dfrac{1}{\ln 5}$.

(ii) Find the probability that it takes more than 3 minutes for Usha to develop a reaction.

(iii) Find the median time for Usha to develop a reaction.

Cambridge International AS & A Level Mathematics
9709 Paper 7 Q7 June 2008

13 The time in minutes taken by candidates to answer a question in an examination has probability density function given by

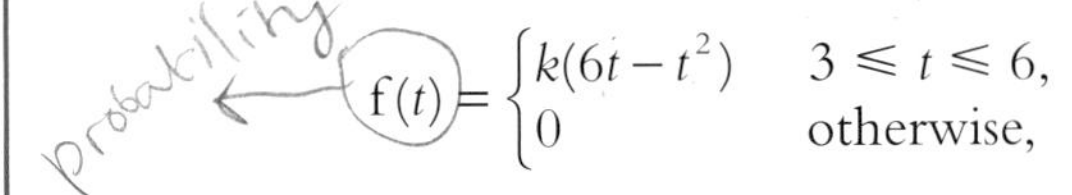

$$f(t) = \begin{cases} k(6t - t^2) & 3 \leqslant t \leqslant 6, \\ 0 & \text{otherwise,} \end{cases}$$

where k is a constant.

(i) Show that $k = \frac{1}{18}$.

(ii) Find the mean time.

(iii) Find the probability that a candidate, chosen at random, takes longer than 5 minutes to answer the question.

(iv) Is the upper quartile of the times greater than 5 minutes, equal to 5 minutes or less than 5 minutes? Give a reason for your answer.

Cambridge International AS & A Level Mathematics
9709 Paper 71 Q5 June 2009

14 The time in hours taken for clothes to dry can be modelled by the continuous random variable with probability density function given by

$$f(t) = \begin{cases} k\sqrt{t} & 1 \leqslant t \leqslant 4, \\ 0 & \text{otherwise,} \end{cases}$$

where k is a constant.

(i) Show that $k = \frac{3}{14}$.

(ii) Find the mean time taken for clothes to dry.

(iii) Find the median time taken for clothes to dry.

(iv) Find the probability that the time taken for clothes to dry is between the mean time and the median time.

Cambridge International AS & A Level Mathematics 9709 Paper 7 Q7 November 2008

15 The random variable T denotes the time in seconds for which a firework burns before exploding. The probability density function of T is given by

$$f(t) = \begin{cases} ke^{0.2t} & 0 \leqslant t \leqslant 5, \\ 0 & \text{otherwise,} \end{cases}$$

where k is a constant.

(i) Show that $k = \frac{1}{5(e-1)}$.

(ii) Sketch the probability density function.

(iii) 80% of fireworks burn for longer than a certain time before they explode. Find this time.

Cambridge International AS & A Level Mathematics 9709 Paper 71 Q5 June 2010

KEY POINTS

1 If X is a continuous random variable with probabilty density function (PDF) $f(x)$

$$\int f(x)\mathrm{d}x = 1$$

$$f(x) \geqslant 0 \quad \text{for all } x$$

$$P(c \leqslant x \leqslant d) = \int_c^d f(x)\mathrm{d}x$$

$$E(X) = \int x f(x)\mathrm{d}x$$

$$\text{Var}(X) = \int x^2 f(x)\mathrm{d}x - [E(X)]^2$$

The mode of X is the value for which $f(x)$ has its greatest magnitude.

2 For the uniform (rectangular) distribution over the interval (a, b)

$$f(x) = \frac{1}{b-a}$$

$$E(X) = \tfrac{1}{2}(a+b)$$

$$\text{Var}(X) = \frac{(b-a)^2}{12}$$

LEARNING OUTCOMES

Now that you have finished this chapter, you should be able to

- use a simple continuous random variable as a model
- understand the meaning of a probability density function (PDF) and be able to use one to find probabilities
- know and use the properties of a PDF
- sketch the graph of a PDF
- find the:
  - mean
  - variance
  - median
  - percentiles

  from a given PDF
- use probability density functions to solve problems.

3 Hypothesis testing using the binomial distribution

You may prove anything by figures.
An anonymous salesman

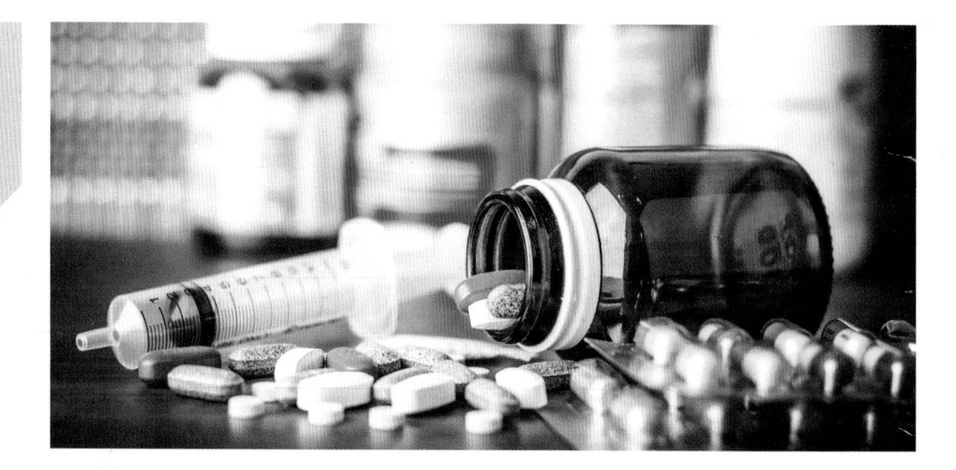

› How can you be certain that any medicine you get from a pharmacy is safe?

Machoman Dan

Just became a father again!
Eight boys in a row – how's that for macho chromosomes?
Even at school I told people I was a real man!

What do you think?

There are two quite different points here.

Maybe you think that Dan is prejudiced, preferring boys to girls. However, you should not let your views on that influence your judgement on the second point, his claim to be biologically different from other people, with special chromosomes.

There are two ways this claim could be investigated: to look at his chromosomes under a high magnification microscope or to consider the statistical evidence. Since you have neither Dan nor a suitable microscope to hand, you must resort to the latter.

If you have eight children you would expect them to be divided about evenly between the sexes, 4 : 4, 5 : 3 or perhaps 6 : 2. When you realise that a baby is on its way you would think it equally likely to be a boy or a girl until it was born, or a scan was carried out, when you would know for certain one way or the other.

In other words, you would say that the probability of it being a boy was 0.5 and that of it being a girl was 0.5. So you can model the number of boys among eight children by the binomial distribution B(8, 0.5).

This gives the probabilities in the table, also shown in Figure 3.1.

Boys	Girls	Probability
0	8	$\frac{1}{256}$
1	7	$\frac{8}{256}$
2	6	$\frac{28}{256}$
3	5	$\frac{56}{256}$
4	4	$\frac{70}{256}$
5	3	$\frac{56}{256}$
6	2	$\frac{28}{256}$
7	1	$\frac{28}{256}$
8	0	$\frac{1}{256}$

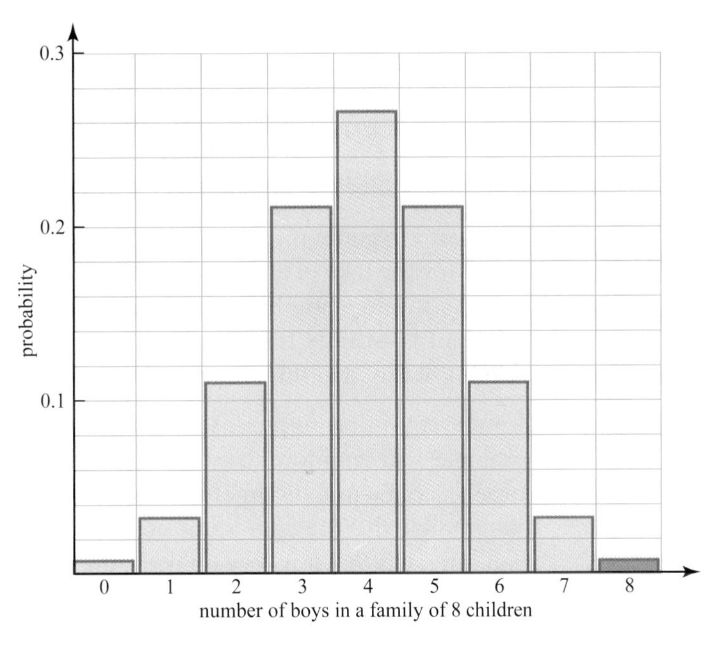

▲ **Figure 3.1**

So you can say that, if a biologically normal man fathers eight children, the probability that they will all be boys is $\frac{1}{256}$ (dark green in Figure 3.1).

This is unlikely but by no means impossible.

Note

The probability of a baby being a boy is not in fact 0.5 but about 0.503. Boys are less tough than girls and so more likely to die in infancy and this seems to be nature's way of compensating. In most societies men have a markedly lower life expectancy as well.

?

› In some countries many people value boys more highly than girls. Medical advances mean that it will soon be possible for parents to decide in advance the sex of their next baby. What would be the effect of this on a country's population if, say, half the parents decided to have only boys and the other half to let nature take its course?

(This is a real problem. The social consequences could be devastating.)

3.1 Defining terms

In the last example we investigated Dan's claim by comparing it to the usual situation, the unexceptional. If we use p for the probability that a child is a boy then the normal state of affairs can be stated as

$$p = 0.5.$$

This is called the **null hypothesis**, denoted by H_0.

Dan's claim (made, he says, before he had any children) was that

$$p > 0.5$$

and this is called the **alternative hypothesis**, H_1.

The word hypothesis (plural **hypotheses**) means a theory that is put forward either for the sake of argument or because it is believed or suspected to be true. An investigation like this is usually conducted in the form of a test, called a **hypothesis test**. There are many different sorts of hypothesis test used in statistics; in this chapter you meet only one of them.

It is never possible to prove something statistically in the sense that, for example, you can prove that the angle sum of a triangle is 180°. Even if you tossed a coin a million times and it came down heads every single time, it is still possible that the coin is unbiased and just happened to land that way. What you can say is that it is very unlikely; the probability of it happening that way is $(0.5)^{1\,000\,000}$, which is a decimal that starts with over 300 000 zeros. This is so tiny that you would feel quite confident in declaring the coin biased.

There comes a point when the probability is so small that you say 'That's good enough for me. I am satisfied that it hasn't happened that way by chance.'

The probability at which you make that decision is called the **significance level** of the test. Significance levels are usually given as percentages; 0.05 is written as 5%, 0.01 as 1%, and so on.

A **test statistic** is a statistic that is determined from the sample data in order to test a hypothesis. In the case of Dan, the test statistic is the number of boys in a sample of eight children.

So for Dan, the question could have been worded:

Test, at the 1% significance level, Dan's claim that his children are more likely to be boys than girls.

The answer would then look like this:

Null hypothesis, H_0:	$p = 0.5$	Boys and girls are equally likely.
Alternative hypothesis, H_1:	$p > 0.5$	Boys are more likely.
Significance level:	1%	

Probability of 8 boys from 8 children $= \frac{1}{256} = 0.0039 = 0.39\%$.

Since $0.39\% < 1\%$ we reject the null hypothesis and accept the alternative hypothesis. We accept Dan's claim.

This example also illustrates some of the problems associated with hypothesis testing. Here is a list of points you should be considering.

3.2 Hypothesis testing checklist

1 Was the test set up before or after the data were known?

The test consists of a null hypothesis, an alternative hypothesis and a significance level.

In this case, the null hypothesis is the natural state of affairs and so does not really need to be stated in advance. Dan's claim 'Even at school I told people I was a real man' could be interpreted as the alternative hypothesis, $p > 0.5$.

The problem is that one suspects that whatever children Dan had he would find an excuse to boast. If they had all been girls, he might have been talking about 'my irresistible attraction for the opposite sex' and if they had been a mixture of girls and boys he would have been claiming 'super-virility' just because he had eight children.

Any test carried out retrospectively must be treated with suspicion.

2 Was the sample involved chosen at random and are the data independent?

The sample was not random and that may have been inevitable. If Dan had lots of children around the country with different mothers, a random sample of eight could have been selected. However, we have no knowledge that this is the case.

The data are the sexes of Dan's children. If there are no multiple births (for example identical twins), then they are independent.

3 Is the statistical procedure actually testing the original claim?

Dan claims to have 'macho chromosomes' whereas the statistical test is of the alternative hypothesis that $p > 0.5$. The two are not necessarily the same. Even if this alternative hypothesis is true, it does not necessarily follow that Dan has macho chromosomes.

The ideal hypothesis test

In the ideal hypothesis test you take the following steps, in this order:

1. Establish the null and alternative hypotheses.
2. Decide on the significance level.
3. Collect suitable data using a random sampling procedure that ensures the items are independent.
4. Conduct the test, doing the necessary calculations.
5. Interpret the result in terms of the original claim, theory or problem.

There are times, however, when you need to carry out a test but it is just not possible to do so as rigorously as this.

If Dan had been a laboratory rat you could have organised that he fathered further babies but this is not possible with a human.

3.3 Choosing the significance level

If, instead of 1%, we had set the significance level at 0.1%, then we would have rejected Dan's claim, since 0.39% > 0.1%. The lower the percentage in the significance level, the more stringent the test.

The significance level you choose for a test involves a balanced judgement.

Imagine that you are testing the rivets on a plane's wing to see if they have lost their strength. Setting a small significance level, say 0.1%, means that you will only declare the rivets weak if you are very confident of your finding. The trouble with requiring such a high level of evidence is that even when they are weak, you may well fail to register the fact, with the possible consequence that the plane crashes. On the other hand if you set a high significance level, such as 10%, you run the risk of declaring the rivets faulty when they are all right, involving the company in expensive and unnecessary maintenance work.

The question of how you choose the best significance level is, however, beyond the scope of this chapter.

Example 3.1

Leonora claims that a die is biased with a tendency to show the number 1. The die was thrown 20 times and the results were as follows.

1	6	6	5	5	1	2	3	2	3
4	4	4	1	4	1	1	4	1	3

Using a 5% significance level, test whether Leonora's claim is correct.

Solution

Let p be the probability of getting 1 on any throw of the die.

Null hypothesis, H_0: $p = \frac{1}{6}$ The die is unbiased.

Alternative hypothesis, H_1: $p > \frac{1}{6}$ The die is biased towards 1.

Significance level: 5%

The results may be summarised as follows.

Score	1	2	3	4	5	6
Frequency	6	2	3	5	2	2

Under the null hypothesis, the number of 1s obtained is modelled by the binomial distribution, $B(20, \frac{1}{6})$ which gives these probabilities:

Number of 1s	Expression	Probability
0	$\left(\frac{5}{6}\right)^{20}$	0.0261
1	$^{20}C_1\left(\frac{5}{6}\right)^{19}\left(\frac{1}{6}\right)$	0.1043
2	$^{20}C_2\left(\frac{5}{6}\right)^{18}\left(\frac{1}{6}\right)^2$	0.1982
3	$^{20}C_3\left(\frac{5}{6}\right)^{17}\left(\frac{1}{6}\right)^3$	0.2379
4	$^{20}C_4\left(\frac{5}{6}\right)^{16}\left(\frac{1}{6}\right)^4$	0.2022
5	$^{20}C_5\left(\frac{5}{6}\right)^{15}\left(\frac{1}{6}\right)^5$	0.1294
6	$^{20}C_6\left(\frac{5}{6}\right)^{14}\left(\frac{1}{6}\right)^6$	0.0647
7	$^{20}C_7\left(\frac{5}{6}\right)^{13}\left(\frac{1}{6}\right)^7$	0.0259
8	$^{20}C_8\left(\frac{5}{6}\right)^{12}\left(\frac{1}{6}\right)^8$	0.0084
⋮	⋮	⋮
20	$\left(\frac{1}{6}\right)^{20}$	0.0000

The probability of 1 coming up between 0 and 5 times is found by adding these probabilities. You get 0.8981 but working to more decimal places and then rounding gives 0.8982, which is correct to 4 decimal places.

If you worked out all these and added them you would get the probability that the number of 1s is 6 or more (up to a possible 20). It is much quicker, however, to find this as 1 – 0.8982 (the answer above) = 0.1018.

Recall from *Probability & Statistics 1* that ^nC_r may also be written as $\binom{n}{r}$. Both notations are used in this book to help you become familiar with both of them.

Calling X the number of 1s occurring when a die is rolled 20 times, the probability of six or more 1s is given by

$$\begin{aligned} P(X \geqslant 6) &= 1 - P(X \leqslant 5) \\ &= 1 - 0.8982 \\ &= 0.1018, \end{aligned}$$

about 10%.

Since 10% > 5%, the null hypothesis (the die is unbiased) is accepted. So Leonora's claim is rejected at the 5% significance level.

The probability of a result at least as extreme as that observed is greater than the 5% cut-off that was set in advance; that is, greater than the chosen significance level.

The alternative hypothesis (the die is biased in favour of the number 1) is rejected, even though the number 1 did come up more often than the other numbers.

❯ Does the procedure in Example 3.1 follow the steps of the ideal hypothesis test?

Note

Notice that this is a test not of the particular result (six 1s) but of a result at least as extreme as this (at least six 1s), the darker area in Figure 3.2. A hypothesis test deals with the probability of an event 'as unusual as or more unusual than' what has occurred.

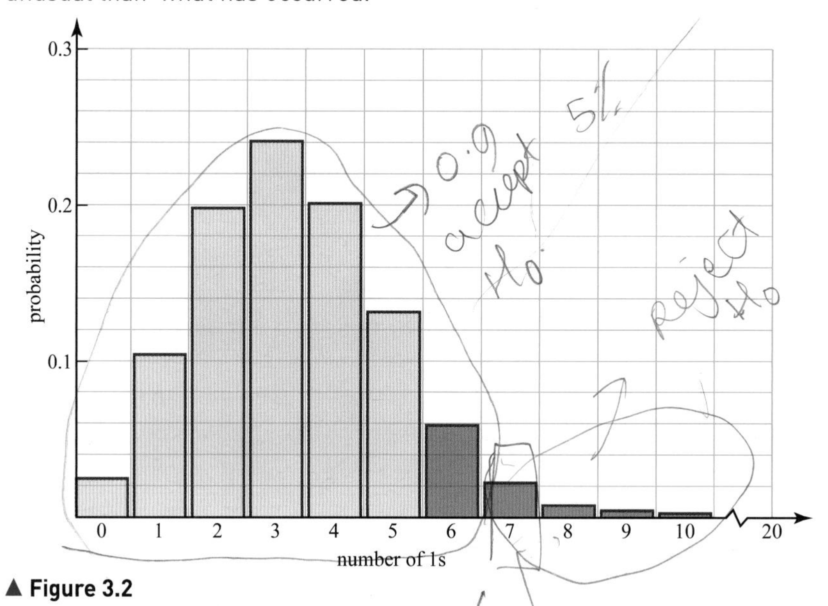

▲ **Figure 3.2**

In the previous example the hypothesis test was based on the binomial distribution, and that is the case for all the tests in this chapter. The conditions for using the binomial distribution are as follows.

- You are conducting trials on a random sample of a certain size, denoted by n.
- On each trial the outcomes may be classified as either **success** or **failure**.

In addition the following modelling assumptions are needed if the binomial distribution is to be a good model and give reliable answers.

- The outcome of each trial is independent of the outcome of any other trial.
- The probability of success is the same on each trial.

You will recognise that these conditions and assumptions are similar to those that you met for the geometric distribution in *Probability and Statistics 1*. The main difference is that in situations modelled by the geometric distribution the trials continue for as long as is necessary to obtain the first success, whereas the binomial distribution is used when the number of trials is known from the outset.

Exercise 3A

In all these questions you should apply this checklist to the hypothesis test.

- Was the test set up before or after the data were known?
- Was the sample used for the test chosen at random and are the data independent?
- Is the statistical procedure actually testing the original claim?

You should also comment critically on whether these steps have been followed.

- Establish the null and alternative hypotheses.
- Decide on the significance level.
- Collect suitable data using a random sampling procedure that ensures the items are independent.
- Conduct the test, doing the necessary calculations.
- Interpret the result in terms of the original claim, theory or problem.

1 Mrs da Silva is running for President. She claims to have 60% of the population supporting her.

She is suspected of overestimating her support and a random sample of 12 people are asked whom they support. Only four say Mrs da Silva.

Test, at the 5% significance level, the hypothesis that she has overestimated her support.

2 A company developed synthetic coffee and claims that coffee drinkers could not distinguish it from the real product. A number of coffee drinkers challenged the company's claim, saying that the synthetic coffee tasted synthetic. In a test, carried out by an independent consumer protection body, 20 people were given a mug of coffee. Ten had the synthetic brand and ten the natural, but they were not told which they had been given.

Out of the ten given the synthetic brand, eight said it was synthetic and two said it was natural. Use this information to test the coffee drinkers' claim (as against the null hypothesis of the company's claim), at the 5% significance level.

M **3** A group of 18 students decides to investigate the truth of the saying that if you drop a piece of toast it is more likely to land butter-side down. They each take one piece of toast, butter it on one side and throw it in the air. Fourteen land butter-side down, the rest butter-side up. Use their results to carry out a hypothesis test at the 1% significance level, stating clearly your null and alternative hypotheses.

M **4** On average 70% of people pass their driving test first time. There are complaints that Mr McTaggart is too harsh and so, unknown to himself, his work is monitored. It is found that he fails six out of ten candidates. Are the complaints justified at the 5% significance level?

M **5** A machine makes bottles. In normal running 5% of the bottles are expected to be cracked, but if the machine needs servicing this proportion will increase. As part of a routine check, 50 bottles are inspected and five are found to be unsatisfactory. Does this provide evidence, at the 5% significance level, that the machine needs servicing?

M **6** An annual mathematics contest contains 15 questions, 5 short and 10 long. The probability that I get a short question right is 0.9. The probability that I get a long question right is 0.5. My performances on questions are independent of each other. Find the probability of the following:

(i) I get all the 5 short questions right.

(ii) I get exactly 8 of the 10 long questions right.

(iii) I get exactly 3 of the short questions and all of the long questions right.

(iv) I get exactly 13 of the 15 questions right.

After some practice, I hope that my performance on the long questions will improve this year. I intend to carry out an appropriate hypothesis test.

(v) State suitable null and alternative hypotheses for the test.

In this year's contest I get exactly 8 of the 10 long questions right.

(vi) Is there sufficient evidence, at the 5% significance level, that my performance on long questions has improved?

7 Isaac claims that 30% of cars in his town are red. His friend Hardip thinks that the proportion is less than 30%. The boys decided to test Isaac's claim at the 5% significance level and found that 2 cars out of a random sample of 18 were red. Carry out the hypothesis test and state your conclusion.

Cambridge International AS & A Level Mathematics 9709 Paper 7 Q1 November 2007

8 Before attending a basketball course, a player found that 60% of his shots made a score. After attending the course the player claimed he had improved. In his next game he tried 12 shots and scored in 10 of them. Assuming shots to be independent, test this claim at the 10% significance level.

Cambridge International AS & A Level Mathematics 9709 Paper 7 Q2 June 2003

3.4 Critical values and critical (rejection) regions

In Example 3.1 the number 1 came up six times and this was not enough for Leonora to show that the die was biased. What was the least number of times 1 would have had to come up for the test to give the opposite result?

We again use X to denote the number of times 1 comes up in the 20 throws and so $X = 6$ means that the number 1 comes up six times.

We know from our earlier work that the probability that $X \leqslant 5$ is 0.8982 and we can use the binomial distribution to work out the probabilities that $X = 6$, $X = 7$, etc.

$$P(X = 6) = {}^{20}C_2\left(\frac{5}{6}\right)^{14}\left(\frac{1}{6}\right)^{6} = 0.0647$$

$$P(X = 7) = {}^{20}C_7\left(\frac{5}{6}\right)^{13}\left(\frac{1}{6}\right)^{7} = 0.0259$$

${}^{20}C_2$ can also be written as $\binom{20}{2}$.

We know $P(X \geqslant 6) = 1 - P(X \leqslant 5) = 1 - 0.8982 = 0.1018$.

0.1018 is a little over 10% and so greater than the significance level of 5%. There is no reason to reject H_0.

What about the case when the number 1 comes up seven times, that is $X = 7$?

$$\begin{aligned} \text{Since} \quad P(X \leqslant 6) &= P(X \leqslant 5) + P(X = 6) \\ P(X \leqslant 6) &= 0.8982 + 0.0647 = 0.9629 \\ \text{So} \quad P(X \geqslant 7) &= 1 - P(X \leqslant 6) \\ &= 1 - 0.9629 = 0.0371 = 3.71\% \end{aligned}$$

Since $3.7\% < 5\%$, H_0 is now rejected in favour of H_1.

You can see that Leonora needed the 1 to come up seven or more times if her claim was to be upheld. She missed by just one. You might think Leonora's 'all or nothing' test was a bit harsh. Sometimes tests are designed so that if the result falls within a certain region further trials are recommended.

In this example the number 7 is the **critical value** (at the 5% significance level), the value at which you change from accepting the null hypothesis to rejecting it. The range of values for which you reject the null hypothesis, in this case $X \geqslant 7$, is called the **critical region** or the **rejection region**.

It is sometimes easier in hypothesis testing to find the critical region and see if your value lies in it, rather than working out the probability of a value at least as extreme as the one you have, the procedure used so far.

The quality control department of a factory tests a random sample of 20 items from each batch produced. A batch is rejected (or perhaps subject to further tests) if the number of faulty items in the sample, X, is more than 2.

This means that the rejection region is $X \geqslant 3$. So the **acceptance region** is $X \leqslant 2$.

It is much simpler for the operator carrying out the test to be told the rejection region (determined in advance by the person designing the procedure) than to have to work out a probability for each test result.

Test procedure

Take 20 pistons

If 3 or more are faulty REJECT the batch

Example 3.2

World-wide 25% of men are colour-blind but it is believed that the condition is less widespread among a group of remote hill tribes. An anthropologist plans to test this by sending field workers to visit villages in that area. In each village 30 men are to be tested for colour-blindness. Find the rejection region for the test at the 5% level of significance.

Solution

Let p be the probability that a man in that area is colour-blind.

Null hypothesis, H_0:	$p = 0.25$	
Alternative hypothesis, H_1:	$p < 0.25$	Less colour-blindness in this area.
Significance level:	5%	

With the hypothesis H_0, if the number of colour-blind men in a sample of 30 is X, then $X \sim B(30, 0.25)$.

The rejection region is the region $X \leqslant k$, where

$$P(X \leqslant k) \leqslant 0.05 \quad \text{and} \quad P(X \leqslant k+1) > 0.05.$$

$$P(X = 0) = (0.75)^{30} = 0.00018$$

$$P(X = 1) = 30(0.75)^{29}(0.25) = 0.00179$$

$$P(X = 2) = \binom{30}{2}(0.75)^{28}(0.25)^2 = 0.00863$$

$\binom{30}{2}$ is the same as ${}^{30}C_2$.

$$P(X = 3) = \binom{30}{3}(0.75)^{27}(0.25)^3 = 0.02685$$

$$P(X = 4) = \binom{30}{4}(0.75)^{26}(0.25)^4 = 0.06042.$$

So $\quad P(X \leqslant 3) = 0.00018 + 0.00179 + 0.00863 + 0.02685 \approx 0.0375 \leqslant 0.05$

but $\quad P(X \leqslant 4) \approx 0.0929 > 0.05.$

Therefore the rejection region is $X \leqslant 3$.

So the acceptance region is $X \geqslant 4$.

› What is the rejection region at the 10% significance level?

In many other hypothesis tests it is usual to find the critical values from tables.

EXPERIMENTS

Mind reading

Here is a simple experiment to see if you can read the mind of a friend whom you know well. The two of you face each other across a table on which is placed a coin. Your friend takes the coin and puts it in one or other hand under the table. You have to guess which one.

Play this game at least 20 times and test at the 10% significance level whether you can read your friend's mind.

Left and right

It is said that if people are following a route that brings them to a T-junction where they have a free choice between turning left and right the majority will turn right.
Design and carry out an experiment to test this hypothesis.

Note

This is taken very seriously by companies choosing stands at exhibitions. It is considered worth paying extra for a location immediately to the right of one of the entrances.

Coloured sweets

Get a large box of coloured sweets and taste the different colours. Choose the colour, C, that you think has the most distinctive flavour. Now close your eyes and get a friend to feed you sweets. Taste each one and say if it is your chosen colour or not. Do this for at least 20 sweets and test at the 10% significance level whether you can pick out those with colour C by taste.

Exercise 3B

You will find it easier to work out the probability that the number not arriving on time is 3, 2, 1 or 0 than to calculate the probability that the number arriving on time is 0, 1, 2, …, 13.

1 A parcel courier firm claims to deliver 90% of its parcels the next day.

A new manager at the firm suspects that the service has recently deteriorated. To test this, 17 parcels are posted and the number of parcels that is delivered by the next day is recorded.

Let p denote the probability that a parcel is delivered the next day.

(i) Write down suitable null and alternative hypotheses for the value of p.

(ii) Write down the

(a) critical region

(b) acceptance region

for the test at the 5% significance level.

(iii) It is found that 13 out of the 17 parcels arrive the next day. Is there enough evidence for the manager to conclude the service has recently deteriorated?

2 Sami has two dice. He claims they are biased towards even numbers. To test his theory, he rolls them eight times so that 16 numbers come up.

Let p denote the probability that the score on one of the dice is even.

(i) Write down suitable null and alternative hypotheses for the value of p.

(ii) Write down the

(a) critical region

(b) acceptance region

for the test at the 5% significance level.

(iii) Of the 16 numbers Sami rolls, 12 are even. Is there enough evidence for Sami to conclude that his dice are biased?

3 Mrs Singh is a maths teacher at Avonford College. She claims that 80% of her students get a grade C or above. Mrs Singh has a class of 18 students.

(i) Find the probability that 17 or more students will achieve a grade C or above if

(a) Mrs Singh's claim is correct

(b) Mrs Singh's claim is incorrect and 82% of her students, on average, achieve a grade C or above.

The Head of Maths thinks the pass rate is higher than 80%. He decides to carry out a hypothesis test at the 10% significance level on Mrs Singh's class of 18 students.

Let p denote the probability that a student passes their maths exam with a grade C or above.

(ii) Write down suitable null and alternative hypotheses for the value of p.

(iii) Write down the critical region for the test.

(iv) Calculate the probability that the Head of Maths will reach the *wrong* conclusion if

(a) Mrs Singh's true pass rate is 80%

(b) Mrs Singh's true pass rate is 82%.

3.5 One-tailed and two-tailed tests

Think back to the two examples in the first part of this chapter.

What would Dan have said if his eight children had all been girls? What would Leonora have said if the number 1 had not come up at all?

In both our examples the claim was not only that something was unusual but that it was so in a particular direction. So we looked only at one side of the distributions when working out the probabilities, as you can see in Figure 3.1 on page 35 and Figure 3.2 on page 40. In both cases we applied **one-tailed tests**. (The word 'tailed' refers to the darker coloured part at the end of the distribution.)

If Dan had just claimed that there was something odd about his chromosomes, then you would have had to work out the probability of a result as extreme on either side of the distribution, in this case eight girls or eight boys, and you would then apply a **two-tailed test**.

Here is an example of a two-tailed test.

Example 3.3

The producer of a television programme claims that it is politically unbiased. 'If you take somebody off the street it is 50 : 50 whether he or she will say the programme favours the government or the opposition', she says.

However, when ten people, selected at random, are asked the question 'Does the programme support the government or the opposition?', nine say it supports the government.

Does this constitute evidence, at the 5% significance level, that the producer's claim is inaccurate?

Solution

Read the last sentence carefully and you will see that it does not say in which direction the bias must be. It does not ask whether the programme is favouring the government or the opposition, only whether the producer's claim is inaccurate. So you must consider both ends of the distribution, working out the probability of such an extreme result either way: 9 or 10 saying it favours the government, or 9 or 10 the opposition. This is a two-tailed test.

If p is the probability that somebody believes the programme supports the government, you have

Null hypothesis, H_0: $p = 0.5$ ← Claim accurate
Alternative hypothesis, H_1: $p \neq 0.5$ ← Claim inaccurate
Significance level: 5%
Two-tailed test so 2.5% in each tail

→

The situation is modelled by the binomial distribution B(10, 0.5) and is shown in Figure 3.3.

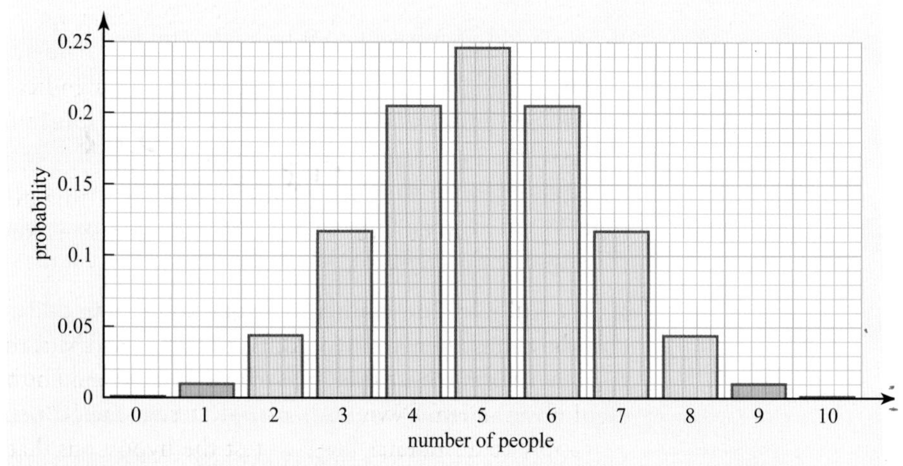

▲ **Figure 3.3**

This gives

$$P(X = 0) = \frac{1}{1024} \qquad P(X = 1) = \frac{10}{1024}$$

$$P(X = 10) = \frac{1}{1024} \qquad P(X = 9) = \frac{10}{1024}$$

where X is the number of people saying the programme favours the government.

Thus the total probability for the two tails is $\frac{22}{1024}$ or 2.15%.

Since 2.15% < 2.5% the null hypothesis is rejected in favour of the alternative, *that the producer's claim is inaccurate.*

Note

You have to look carefully at the way a test is worded to decide if it should be one-tailed or two-tailed.

Dan claimed his chromosomes made him more likely to father boys than girls. That requires a one-tailed test.

Leonora claimed the die was biased in the direction of too many 1s. Again a one-tailed test.

The test of the television producer's claim was for inaccuracy in either direction and so a two-tailed test was needed.

Exercise 3C

1 To test the claim that a coin is biased, it is tossed 12 times. It comes down heads three times. Test at the 10% significance level whether this claim is justified.

2 A biologist discovers a colony of a previously unknown type of bird nesting in a cave. Out of the 16 chicks that hatch during his period of investigation, 13 are female. Test at the 5% significance level whether this supports the view that the sex ratio for the chicks differs from 1:1.

3 People entering an exhibition have to choose whether to turn left or right. Out of the first 12 people, nine turn left and three right. Test at the 5% significance level whether people are more likely to turn one way than the other.

4 A multiple choice test has 15 questions, with the answer for each allowing five options, A, B, C, D and E. All the students in a class tell their teacher that they guessed all 15 answers. The teacher does not believe them. Devise a two-tailed test at the 10% significance level to apply to a student's mark to test the hypothesis that the answers were not selected at random.

5 When a certain language is written down, 15% of the letters are Z. Use this information to devise a test at the 10% significance level that somebody who does not know the language could apply to a short passage, 50 letters long, to determine whether it is written in the same language.

6 A seed firm states on a packet of rare seeds that the germination rate is 20%. The packet contains 25 seeds.

(i) How many seeds would you expect to germinate out of the packet?

(ii) What is the probability of exactly five seeds germinating?

A man buys a packet and only one seed germinates.

(iii) Is he justified in complaining?

3.6 Type I and Type II errors

There are two types of error that can occur when a hypothesis test is carried out. They are illustrated in the following example.

Example 3.4

A gold coin is used for the toss at a country's football matches but it is suspected of being biased. It is suggested that it shows heads more often than it should. A test is planned in which the coin is to be tossed 19 times and the results recorded. It is decided to use a 5% significance level; so, if the coin shows heads 14 or more times, it will be declared biased.

What errors are possible in interpreting the test result? →

Solution

Two types of error are possible.

A Type I error

In this case the coin is actually unbiased, so the probability, p, of it showing heads is given by $p = 0.5$. However, it happens to come up heads 14 or more times and so is incorrectly declared to be biased.

The probabilities of possible outcomes from 19 tosses when $p = 0.5$ can be found using the binomial distribution. Some of them are given, to 2 significant figures, in the table below.

Number of heads	$\geqslant 10$	$\geqslant 11$	$\geqslant 12$	$\geqslant 13$	$\geqslant 14$	$\geqslant 15$	$\geqslant 16$
Probability	0.50	0.32	0.18	0.084	0.032	0.010	0.0022

The table shows that the probability of getting 14 or more heads, and so making the error of rejecting the true null hypothesis that $p = 0.5$, is 0.032 and so just less than the 5% significance level. This type of error, where a null hypothesis is rejected despite being correct, is called a Type I error. The figures in the table illustrate the fact that for a binomial test the probability of making a Type I error is either equal to the significance level of the test or slightly less than it. For most other hypothesis tests it is equal to the significance level; indeed, that is the meaning of the term 'significance level', the probability of rejecting a true null hypothesis.

A Type II error

The other type of error occurs when the null hypothesis is in fact false but is nonetheless accepted. Imagine that the gold coin is actually biased with $p = 0.8$ and that it shows heads 12 times. In this test, the null hypothesis is rejected if the number of heads is 14 or more, and so it is accepted if the number of heads is less than 14.

Since $12 < 14$, the null hypothesis is accepted, even though it is in fact false. This is called a Type II error, where a false null hypothesis is accepted.

In this case, it is possible to use the binomial distribution to work out the probability of a Type II error. When $p = 0.8$, the probability that when the coin is tossed 19 times the number of heads is less than 14 can be found to be 0.163, and so this is the probability of a Type II error in this example.

Note

1. Notice that it was only possible to find the probability of a Type II error in Example 3.4 because the value of the population parameter under consideration was known: $p = 0.8$. Since finding out about this parameter is the object of the test, it would be unusual for it to be known. So, in practice, it is often not possible to calculate the probability of a Type II error. By contrast, no calculation at all is needed to find the probability of a Type I error; it is the significance level of the test.
2. For a given sample size, the probabilities of the two types of errors are linked. In Example 3.4, the probability of a Type II error could be reduced by making the test more severe; instead of requiring 14 or more heads to declare the coin biased, it could be reduced to 13 or perhaps 12. However, that would increase the probability of a Type I error.
3. The circumstances under which these errors occur is shown below.

		Decision	
		Accept H_0 (decide the coin is unbiased)	**Reject H_0 (decide the coin is biased)**
Reality	**The null hypothesis, H_0, is true.**	Correct decision	H_0 wrongly rejected: Type I error
	The null hypothesis, H_0, is false.	H_0 wrongly accepted: Type II error	Correct decision

In summary:

- A Type I error occurs when the sample leads you to wrongly reject H_0 when it is in fact true.
- A Type II error occurs when the sample leads you to wrongly accept H_0 when it is in fact false.

Example 3.5

It is known that 60% of the moths of a certain species are red; the rest are yellow. A biologist finds a new colony of these moths and observes that more of them seem to be red than she would expect. She designs an experiment in which she will catch 10 moths at random, observe their colour and then release them. She will then carry out a hypothesis test using a 5% significance level.

(i) State the null and alternative hypotheses for this test.

(ii) Find the rejection region.

(iii) Find the probability of a Type I error.

(iv) If in fact the proportion of red moths is 80%, find the probability that the test will result in a Type II error.

Solution

(i) Let p be the probability that a randomly selected moth is red.

Null hypothesis: $H_0: p = 0.6$ The proportion of red moths in this colony is 60%.

Alternative hypothesis: $H_1: p > 0.6$ The proportion of red moths is greater than 60%.

(ii) Assuming H_0 is true, you can calculate the following probabilities for the 10 moths in the sample.

All 10 moths are red: $(0.6)^{10} = 0.0060...$

9 are red and 1 yellow: ${}^{10}C_1 \times (0.6)^9 \times 0.4 = 0.0403...$

8 are red and 2 yellow: ${}^{10}C_2 \times (0.6)^8 \times (0.4)^2 = 0.1209...$

There is no need to go any further.

The probability that there are nine or ten red moths is

$$0.0403... + 0.0060... = 0.0463...$$

and this is less than the 5% significance level.

The probability that there are eight, nine or ten red moths is

$$0.1209... + 0.0403... + 0.0060... = 0.167...$$

and this is greater than 5%.

So the rejection region for this test is 9 or 10 red moths.

> So the acceptance region is 8 or fewer red moths.

(iii) A Type I error occurs when a true null hypothesis is rejected.

In this case if H_0 is true, and so $p = 0.6$, the probability of it being rejected because a particular sample has 9 or 10 red moths has already been worked out to be 0.0463... in part (ii). When rounded to 3 significant figures, this gives 0.0464.

So the probability of a Type 1 error is 0.0464 (to 3 s.f.).

(iv) If the proportion of red moths is 80%, the correct result from the test would be for the null hypothesis to be rejected in favour of the alternative hypothesis. The probability of this happening is

$${}^{10}C_1 \times (0.8)^9 \times 0.2 + (0.8)^{10} = 0.376 \text{ (to 3 s.f.)}$$

A Type II error occurs when this result does not occur.

So in this situation the probability of a Type II error is $1 - 0.376 = 0.624$.

Exercise 3D

1 At a certain airport 20% of people take longer than an hour to check in. A new computer system is installed, and it is claimed that this will reduce the time to check in. It is decided to accept the claim if, from a random sample of 22 people, the number taking longer than an hour to check in is either 0 or 1.

(i) Calculate the significance level of the test.

(ii) State the probability that a Type I error occurs.

(iii) Calculate the probability that a Type II error occurs if the probability that a person takes longer than an hour to check in is now 0.09.

Cambridge International AS & A Level Mathematics
9709 Paper 7 Q4 June 2007

2 A manufacturer claims that 20% of sugar-coated chocolate beans are red. George suspects that this percentage is actually less than 20% and so he takes a random sample of 15 chocolate beans and performs a hypothesis test with the null hypothesis $p = 0.2$ against the alternative hypothesis $p < 0.2$. He decides to reject the null hypothesis in favour of the alternative hypothesis if there are 0 or 1 red beans in the sample.

(i) With reference to this situation, explain what is meant by a Type I error.

(ii) Find the probability of a Type I error in George's test.

Cambridge International AS & A Level Mathematics
9709 Paper 7 Q2 November 2005

3 In a certain city it is necessary to pass a driving test in order to be allowed to drive a car. The probability of passing the driving test at the first attempt is 0.36 on average. A particular driving instructor claims that the probability of his pupils passing at the first attempt is higher than 0.36. A random sample of 8 of his pupils showed that 7 passed at the first attempt.

(i) Carry out an appropriate hypothesis test to test the driving instructor's claim, using a significance level of 5%.

(ii) In fact, most of this random sample happened to be careful and sensible drivers. State which type of error in the hypothesis test (Type I or Type II) could have been made in these circumstances and find the probability of this type of error when a sample of size 8 is used for the test.

Cambridge International AS & A Level Mathematics
9709 Paper 71 Q4 June 2009

4 It is claimed that a certain 6-sided die is biased so that it is more likely to show a six than if it was fair. In order to test this claim at the 10% significance level, the die is thrown 10 times and the number of sixes is noted.

(i) Given that the die shows a six on 3 of the 10 throws, carry out the test.

On another occasion the same test is carried out again.

(ii) Find the probability of a Type I error.

(iii) Explain what is meant by a Type II error in this context.

Cambridge International AS & A Level Mathematics 9709 Paper 71 Q6 November 2010

5 A cereal manufacturer claims that 25% of cereal packets contain a free gift. Lola suspects that the true proportion is less than 25%. In order to test the manufacturer's claim at the 5% significance level, she checks a random sample of 20 packets.

(i) Find the critical region for the test.

(ii) Hence find the probability of a Type I error.

Lola finds that 2 packets in her sample contain a free gift.

(iii) State, with a reason, the conclusion she should draw.

Cambridge International AS & A Level Mathematics 9709 Paper 71 Q4 November 2012

6 At the last election, 70% of people in Apoli supported the president. Luigi believes that the same proportion support the president now. Maria believes that the proportion who support the president now is 35%. In order to test who is right, they agree on a hypothesis test, taking Luigi's belief as the null hypothesis. They will ask 6 people from Apoli, chosen at random, and if more than 3 support the president they will accept Luigi's belief.

(i) Calculate the probability of a Type I error.

(ii) If Maria's belief is true, calculate the probability of a Type II error.

(iii) In fact 2 of the 6 people say that they support the president. State which error, Type I or Type II, might be made. Explain your answer.

Cambridge International AS & A Level Mathematics 9709 Paper 71 Q6 November 2013

KEY POINTS

1. **Hypothesis testing checklist**
   - Was the test set up before or after the data were known?
   - Was the sample involved chosen at random and are the data independent?
   - Is the statistical procedure actually testing the original claim?
2. **Steps for conducting a hypothesis test**
   - Establish the null and alternative hypotheses.
   - Decide on the significance level.
   - Collect suitable data using a random sampling procedure that ensures the items are independent.
   - Conduct the test, doing the necessary calculations.
   - Interpret the result in terms of the original claim, theory or problem.
3. A Type I error occurs when a true null hypothesis is rejected. The probability of a Type I error occurring is less than or equal to the significance level of the test.
4. A Type II error occurs when a false null hypothesis is accepted. The probability of a Type II error occurring depends on the (unknown) value of the population parameter; in a binomial test the parameter is p.

LEARNING OUTCOMES

Now that you have finished this chapter, you should be able to

- understand the following terms:
  - hypothesis test
  - null hypothesis
  - alternative hypothesis
  - test statistic
  - rejection region (or critical region)
  - acceptance region
  - significance level
- formulate hypotheses and carry out a hypothesis test, including:
  - understanding the difference between one-tailed and two-tailed tests
- understand Type I errors and Type II errors
- calculate the probabilities of making Type I and Type II errors.

4 Hypothesis testing and confidence intervals using the normal distribution

When we spend money on testing an item, we are buying confidence in its performance.
Anthony Cutler (1954–)

4.1 Interpreting sample data using the normal distribution

Sydney set to become greenhouse?

from our Science Correspondent Ama Williams

On a recent visit to a college in Sydney, I was intrigued to find experiments being conducted to measure the level of carbon dioxide in the air we are all breathing. Readers will of course know that high levels of carbon dioxide are associated with the greenhouse effect.

Lecturer Ray Peng showed me round his laboratory. 'It is delicate work, measuring parts per million, but I am trying to establish what is the normal level in this area. Yesterday we took ten readings and you can see the results for yourself: 336, 334, 332, 332, 331, 331, 330, 330, 328, 326.'

When I commented that there seemed to be a lot of variation between the readings, Ray assured me that that was quite in order.

'I have taken hundreds of these measurements in the past,' he said. 'There is always a standard deviation of 2.5. That's just natural variation.'

I suggested to Ray that his students should test whether these results are significantly above the accepted value of 328 parts per million. Certainly they made me feel uneasy. Is the greenhouse effect starting here in Australia?

Ray Peng has been trying to establish the carbon dioxide level in Sydney.

- How do you interpret his figures?
- Do you think the correspondent has a point when she says she is worried that the greenhouse effect is already happening in Australia?

If suitable sampling procedures have not been used, then the resulting data may be worthless, indeed positively misleading. You may wonder if that is the case with Ray's figures, and about the accuracy of his analysis of the samples too. His data are used in subsequent working in this chapter, but you may well feel there is something of a question mark hanging over them. You should always be prepared to treat data with a healthy degree of caution.

Putting aside any concerns about the quality of the data, what conclusions can you draw from them?

Estimating the population mean, μ

Ray's data were as follows.

336 334 332 332 331 331 330 330 328 326

His intention in collecting them was to estimate the mean of the parent population, the **population mean**.

The mean of these figures, the **sample mean**, is given by

$$\bar{x} = \frac{336 + 334 + 332 + 332 + 331 + 331 + 330 + 330 + 328 + 326}{10}$$

$$= 331.$$

What does this tell you about the population mean, μ?

It tells you that it is about 331 but it certainly does not tell you that it is definitely and exactly 331. If Ray took another sample, its mean would probably not be 331 but you would be surprised (and suspicious) if it were very far away from it. If he took lots of samples, all of size 10, you would expect their means to be close together but certainly not all the same.

If you took 1000 such samples, each of size 10, the distribution of their means might look like Figure 4.1. You will notice that this distribution looks rather like the normal distribution and so may well wonder if this is indeed the case.

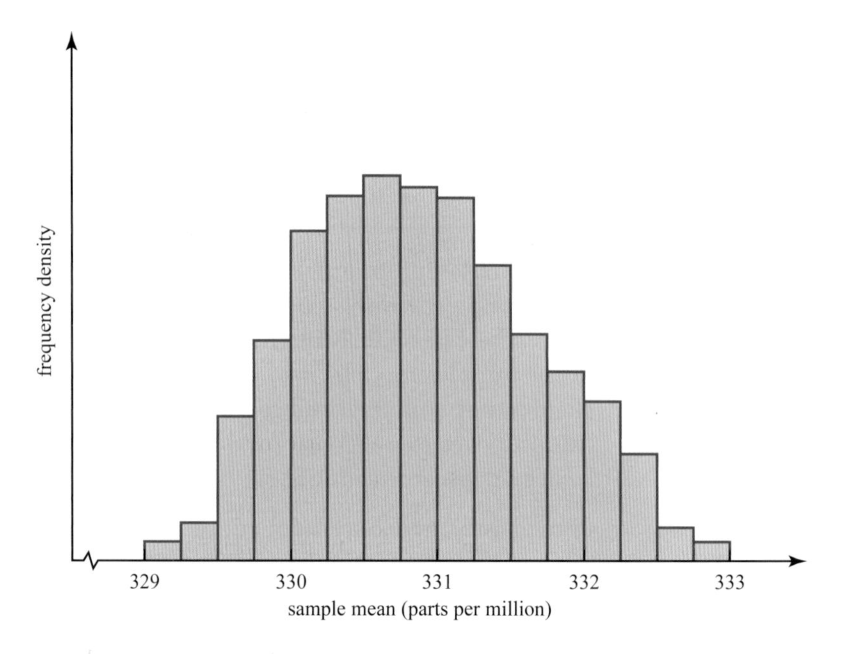

▲ **Figure 4.1**

The distribution of sample means

In this chapter, it is assumed that the underlying population has a normal distribution with mean μ and standard deviation σ so it can be denoted by $N(\mu, \sigma^2)$. In that case the distribution of the means of samples of size n is indeed normal; its mean is μ and its standard deviation is $\frac{\sigma}{\sqrt{n}}$. This is called the **sampling distribution of the means**, or often just the **sampling distribution**, and is denoted by $N\left(\mu, \frac{\sigma^2}{n}\right)$. This is illustrated in Figure 4.2. It is a special case of the Central Limit Theorem, which you will meet later in this chapter, on page 69.

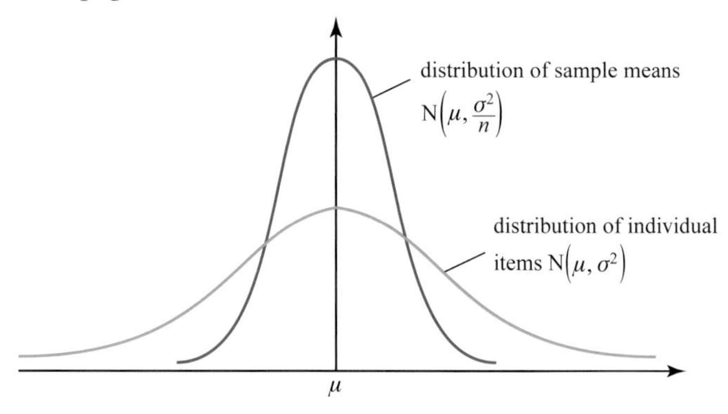

▲ **Figure 4.2**

A hypothesis test for the mean using the normal distribution

If your intention in collecting sample data is to test a theory, then you should set up a hypothesis test.

Ray Peng was mainly interested in establishing data on carbon dioxide levels for Sydney. The correspondent, however, wanted to know whether levels were above normal, and so she could have set up and conducted a test, such as the one in Example 4.1 below.

Example 4.1

Ama Williams believes that the carbon dioxide level in Sydney has risen above the usual level of 328 parts per million. A sample of 10 specimens of Sydney air is collected and the carbon dioxide level within the specimens is determined. The results are as follows.

336 334 332 332 331 331 330 330 328 326

Extensive previous research has shown that the standard deviation of the levels within such samples is 2.5, and that the distribution may be assumed to be normal.

Use these data to test, at the 0.1% significance level, Ama's belief that the level of carbon dioxide at Sydney is above normal.

Solution

As usual with hypothesis tests, you use the distribution of the statistic you are measuring, in this case the normal distribution of the sample means, to decide which values of the test statistic are sufficiently extreme as to suggest that the alternative hypothesis, not the null hypothesis, is true.

Null hypothesis, H_0:	$\mu = 328$	The level of carbon dioxide in Sydney is normal.
Alternative hypothesis, H_1:	$\mu > 328$	The level of carbon dioxide in Sydney is above normal.

One-tailed test

The significance level is 0.1%. This is the probability of a Type I error for this test.

Method 1: Using critical regions

Since the distribution of sample means is $N\left(\mu, \frac{\sigma^2}{n}\right)$, critical values for a test on the sample mean are given by $\mu \pm k \times \frac{\sigma}{\sqrt{n}}$.

In this case, if H_0 is true, $\mu = 328$; $\sigma = 2.5$; $n = 10$. →

The test is one-tailed, for $\mu > 328$, so only the right-hand tail applies. This gives a value of $k = 3.090$ since normal distribution tables give $\Phi(3.090) = 0.999$ and so $1 - \Phi(3.090) = 0.001$.

The critical value is thus $328 + 3.09 \times \dfrac{2.5}{\sqrt{10}} = 330.4$, as shown in Figure 4.3.

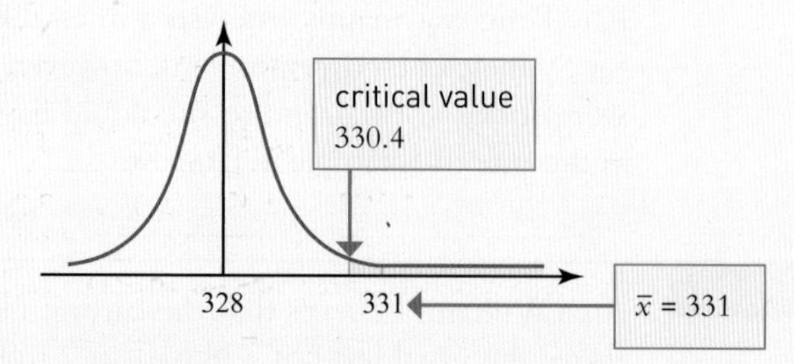

▲ **Figure 4.3**

However, the sample mean $\bar{x} = 331$, and $331 > 330.4$.

Therefore the sample mean lies within the critical region, and so the null hypothesis is rejected in favour of the alternative hypothesis: that the mean carbon dioxide level is above 328, at the 0.1% significance level.

Method 2: Using probabilities

The distribution of sample means, $\bar{X}$, is $\mathrm{N}\left(\mu, \dfrac{\sigma^2}{n}\right)$.

According to the null hypothesis, $\mu = 328$ and it is known that $\sigma = 2.5$ and $n = 10$.

So this distribution is $\mathrm{N}\left(328, \dfrac{2.5^2}{10}\right)$; see Figure 4.4.

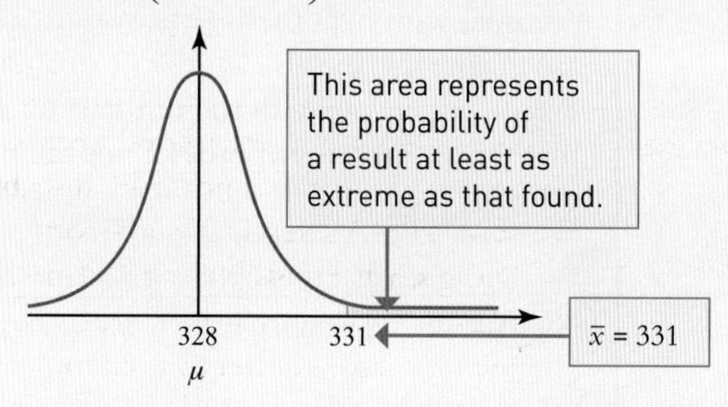

▲ **Figure 4.4**

The probability of the mean, $\bar{X}$, of a randomly chosen sample being greater than the value found, i.e. 331, is given by

$$\mathrm{P}(\bar{X} \geqslant 331) = 1 - \Phi\left(\frac{331 - 328}{\frac{2.5}{\sqrt{10}}}\right)$$

$$= 1 - \Phi(3.79)$$

$$= 1 - 0.999\,93 = 0.000\,07$$

The figure 0.99993 comes from normal distribution tables for suitable values of z.

Since $0.00007 < 0.001$, the required significance level (0.1%), the null hypothesis is rejected in favour of the alternative hypothesis.

You should be aware that values given in tables are rounded. Consequently the final digit of an answer you obtain using tables may not be quite the same as if you have used the statistical function on your calculator instead. Most tables give 4 or 5 figures so it is good practice to round your final answer to 3 significant figures.

Method 3: Using critical ratios

The **critical ratio** is given by $z = \dfrac{\text{observed value} - \text{expected value}}{\text{standard deviation}}$.

In this case $z = \dfrac{331 - 328}{\frac{2.5}{\sqrt{10}}} = 3.79$

This is now compared with the critical value for z given in your tables.

p	0.75	0.90	0.95	0.975	0.99	0.995	0.9975	0.999	0.9995
z	0.674	1.282	1.645	1.960	2.326	2.576	2.807	3.090	3.291

▲ **Figure 4.5 Critical values for the normal distribution**

So the critical value is $z = 3.090$.

Since $3.79 > 3.09$, H_0 is rejected.

Note

1. A hypothesis test should be formulated before the data are collected and not after. If sample data lead you to form a hypothesis, then you should plan a suitable test and collect further data on which to conduct it. It is not clear whether or not the test in the previous example was being carried out on the same data that were used to formulate the hypothesis.
2. If the data were not collected properly, any test carried out on them may be worthless.

Example 4.2

Observations over a long period of time have shown that the mass of adult males of a type of bat is normally distributed with mean 110 g and standard deviation 10 g. A scientist has a theory that in one area these bats are becoming smaller, possibly as an adaptation to changes in their environment. He plans to trap 20 adult male bats, weigh them and then release them. He will then use the data to carry out a suitable hypothesis test at the 5% significance level.

(i) State the null and alternative hypotheses.

(ii) Find the critical value for the test.

(iii) Find the probability of a Type I error.

In fact, the mean mass of the bats has reduced to 108 g but the standard deviation has remained unaltered.

(iv) Calculate the probability that the test will produce a Type II error.

The mean mass of the scientist's sample of bats is 107 g.

(v) Carry out the hypothesis test and state what type of error, if any, results.

→

Solution

(i) The hypotheses are:

Null hypothesis H_0:	$\mu = 110$	The mean mass of the bats is still 110 g.
Alternative hypothesis H_1:	$\mu < 110$	The mean mass of the bats is less than 110 g.

(ii) This is a one-tailed test at the 5% significance level so the critical value is:

$$\bar{X} = 110 - 1.645 \times \frac{10}{\sqrt{20}} = 106.3 \text{ to 1 d.p.}$$

where $\bar{X}$ is the sample mean.

The null hypothesis will be rejected if $\bar{X} < 106.3$.

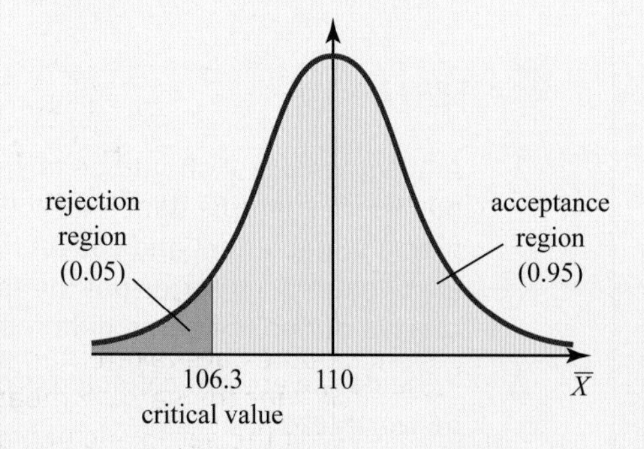

▲ **Figure 4.6**

(iii) A Type I error occurs when a true null hypothesis is rejected. In this case, the probability of this happening is represented by the dark pink area in Figure 4.6. It is just the same as the significance level of the test and so is 5% or 0.05.

(iv) A Type II error will occur if $\bar{X} > 106.3$ because in that case the null hypothesis, which is false, will be accepted.

In fact $\mu = 108$ and so the probability that $\bar{X} > 106.3$ is given by

$$1 - \Phi\left(\frac{106.3 - 108}{\frac{10}{\sqrt{20}}}\right) = 0.77 \text{ (to 2 s.f.).}$$

(v) Since $107 > 106.32$, the null hypothesis is accepted. The evidence does not support the scientist's theory.

However, this is the wrong result so a Type II error has occurred. The answer from part (iv) shows that with the test set up as it was, a Type II error is quite likely to occur.

Known and estimated standard deviation

Notice that you can only use this method of hypothesis testing if you already know the value of the standard deviation of the parent population, σ. Ray Peng had said that from taking hundreds of measurements he knew it to be 2.5.

It is more often the situation that you do not know the population standard deviation or variance and so have to estimate them from your sample data. The estimated standard deviation, s, is worked out using slightly different formulae from those you met in *Probability & Statistics 1*. In certain places $n - 1$ is used instead of n.

To calculate an unbiased estimate of the population mean and variance from a sample you should use the following formulae:

Estimated mean, $\bar{x} = \dfrac{\sum x}{n}$

Estimated variance, $s^2 = \dfrac{1}{n-1}\left(\sum x^2 - \dfrac{\left(\sum x\right)^2}{n}\right)$

An alternative notation to s for the estimated standard deviation is $\hat{\sigma}$.

Note

If repeated samples are taken from a certain population, then slightly different results for the sample mean and variance might be obtained each time. However, if the sample is unbiased, you can work out an unbiased estimate of the mean and variance of the population from which the sample was taken.

Example 4.3

An IQ test, established some years ago, was designed to have a mean score of 100. A researcher puts forward a theory that people are becoming more intelligent (as measured by this particular test). She selects a random sample of 150 people, all of whom take the test. The results of the tests, where x represents the score obtained, are $n = 150$, $\Sigma x = 15\,483$, $\Sigma x^2 = 1\,631\,680$.

Carry out a suitable hypothesis test on the researcher's theory, at the 1% significance level. You may assume that the test scores are normally distributed.

Solution

H_0: The parent population mean is unchanged, i.e. $\mu = 100$.

H_1: The parent population mean has increased, i.e. $\mu > 100$.

One-tailed test

The significance level is 1%. This is the probability of Type I error for this test.

➜

From the sample, unbiased estimates for the mean and standard deviation are:

$$\bar{x} = \frac{\sum x}{n} = \frac{15\,483}{150} = 103.22$$

$$s^2 = \frac{1}{n-1}\left(\sum x^2 - \frac{\left(\sum x\right)^2}{n}\right) = \frac{1}{149}\left(1\,631\,680 - \frac{15\,483^2}{150}\right) = 224.998\ldots$$

So $s = 15.0$ (to 3 s.f.).

The standardised z value corresponding to $\bar{x} = 103.22$ is calculated using $\mu = 100$ and approximating σ by $s = 15.0$.

$$z = \frac{\bar{x} - \mu}{\frac{\sigma}{\sqrt{n}}} = \frac{103.22 - 100}{\frac{15}{\sqrt{150}}} = 2.629$$

For the 1% significance level, the critical value is $z = 2.326$.

The test statistic is compared with the critical value and since 2.629 > 2.326 the null hypothesis is rejected.

The evidence supports the view that scores on this IQ test are now higher; see Figure 4.7.

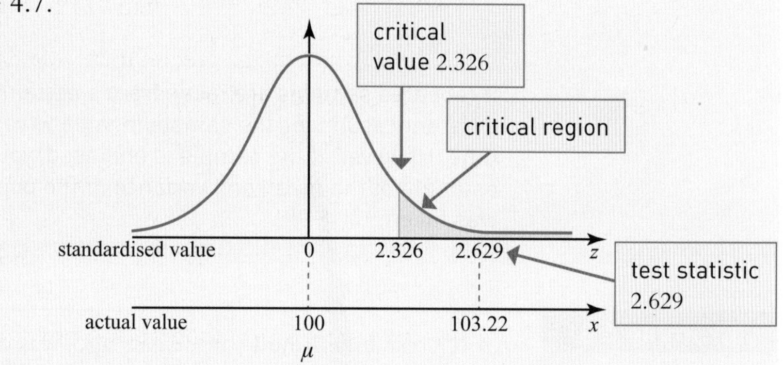

▲ **Figure 4.7**

Exercise 4A

1 For each of the following, the random variable $X \sim N(\mu, \sigma^2)$, with *known* standard deviation. A random sample of size n is taken from the parent population and the sample mean, $\bar{x}$, is calculated.

Carry out hypothesis tests, given H_0 and H_1, at the significance level indicated.

	σ	n	$\bar{x}$	H_0	H_1	Sig. level
(i)	8	6	195	$\mu = 190$	$\mu > 190$	5%
(ii)	10	10	47.5	$\mu = 55$	$\mu < 55$	1%
(iii)	15	25	104.7	$\mu = 100$	$\mu \neq 100$	10%
(iv)	4.3	15	34.5	$\mu = 32$	$\mu > 32$	2%
(v)	40	12	345	$\mu = 370$	$\mu \neq 370$	5%

2 A machine is designed to make paperclips with mean mass 4.00 g and standard deviation 0.08 g. The distribution of the masses of the paperclips is normal. Find

(i) the probability that an individual paperclip, chosen at random, has mass greater than 4.04 g

(ii) the value of $\frac{\sigma}{\sqrt{n}}$ for random samples of 25 paperclips

(iii) the probability that the mean mass of a random sample of 25 paperclips is greater than 4.04 g.

A quality control officer weighs a random sample of 25 paperclips and finds their total mass to be 101.2 g.

(iv) Conduct a hypothesis test at the 5% significance level of whether this provides evidence of an increase in the mean mass of the paperclips. State your null and alternative hypotheses clearly.

CP **3** It is known that the mass of a certain type of lizard has a normal distribution with mean 72.7 g and standard deviation 4.8 g. A zoologist finds a colony of lizards in a remote place and is not sure whether they are of the same type. In order to test this, she collects a sample of 12 lizards and weighs them, with the following results.

80.4 67.2 74.9 78.8 76.5 75.5 80.2 81.9 79.3 70.0 69.2 69.1

(i) Write down, in precise form, the zoologist's null and alternative hypotheses, and state whether a one-tailed or two-tailed test is appropriate.

(ii) Carry out the test at the 5% significance level and write down your conclusion.

(iii) Would your conclusion have been the same at the 10% significance level?

CP **4** Observations over a long period of time have shown that the midday temperature at a particular place during the month of June is normally distributed with a mean value of 23.9 °C with standard deviation 2.3 °C. An ecologist sets up an experiment to collect data for a hypothesis test of whether the climate is getting hotter. She selects at random 20 June days over a five-year period and records the midday temperature. Her results (in °C) are as follows.

20.1	26.2	23.3	28.9	30.4	28.4	17.3	22.7	25.1	24.2
15.4	26.3	19.3	24.0	19.9	30.3	32.1	26.7	27.6	23.1

(i) State the null and alternative hypotheses that the ecologist should use.

(ii) Carry out the test at the 10% significance level and state the conclusion.

(iii) Calculate an unbiased estimate of the population variance and comment on it.

CP **5** The keepers of a lighthouse were required to keep records of weather conditions. Analysis of their data from many years showed the visibility at midday to have a mean value of 14 nautical miles with standard deviation 5.4 nautical miles. A new keeper decided he would test his theory that the air had become less clear (and so visibility reduced) by carrying out a hypothesis test on data collected for his first 36 days on duty. His figures (in nautical miles) were as follows.

35	21	12	7	2	1.5	1.5	1	0.25	0.25	15	17
18	20	16	11	8	8	9	17	35	35	4	0.25
0.25	5	11	28	35	35	16	2	1	0.5	0.5	1

(i) Write down a distributional assumption for the test to be valid.

(ii) Write down suitable null and alternative hypotheses.

(iii) Carry out the test at the 2.5% significance level and state the conclusion that the lighthouse keeper would have come to.

(iv) Criticise the sampling procedure used by the keeper and suggest a better one.

CP **6** The packaging on a type of electric light bulb states that the average lifetime of the bulbs is 1000 hours. A consumer association thinks that this is an overestimate and tests a random sample of 64 bulbs, recording the lifetime, x hours, of each bulb. You may assume that the distribution of the bulbs' lifetimes is normal.

The results are summarised as follows.

$$n = 64, \qquad \Sigma x = 63910.4, \qquad \Sigma x^2 = 63824061$$

(i) Calculate unbiased estimates for the population mean and variance.

(ii) State suitable null and alternative hypotheses to test whether the statement on the packaging is overestimating the lifetime of this type of bulb.

(iii) Carry out the test, at the 5% significance level, stating your conclusions carefully.

7 A sample of 40 observations from a normal distribution X gave $\Sigma x = 24$ and $\Sigma x^2 = 596$. Performing a two-tailed test at the 5% level, test whether the mean of the distribution is zero.

CP **8** A random sample of 75 eleven-year-olds performed a simple task and the time taken, t minutes, was noted for each. You may assume that the distribution of these times is normal.

The results are summarised as follows.

$$n = 75, \qquad \Sigma t = 1215, \qquad \Sigma t^2 = 21708$$

(i) Calculate unbiased estimates for the population mean and variance.

(ii) State suitable null and alternative hypotheses to test whether there is evidence that the mean time taken to perform this task is greater than 15 minutes.

(iii) Carry out the test, at the 1% significance level, stating your conclusions carefully.

CP **9** Bags of sugar are supposed to contain, on average, 2 kg of sugar. A quality controller suspects that they actually contain less than this amount, and so 90 bags are taken at random and the mass, x kg, of sugar in each is measured. You may assume that the distribution of these masses is normal.

The results are summarised as follows.

$$n = 90, \qquad \Sigma x = 177.9, \qquad \Sigma x^2 = 353.1916$$

(i) Calculate unbiased estimates for the population mean and variance.

(ii) State suitable null and alternative hypotheses to test whether there is any evidence that the sugar is being sold 'underweight'.

(iii) Carry out the test, at the 2% significance level, stating your conclusions carefully.

CP **10** A machine produces jars of skin cream, filled to a nominal volume of 100 ml. The machine is actually supposed to be set to 105 ml, to ensure that most jars actually contain more than the nominal volume of 100 ml. You may assume that the distribution of the volume of skin cream in a jar is normal.

To check that the machine is correctly set, 80 jars are chosen at random, and the volume, x ml, of skin cream in each is measured.

The results are summarised as follows.

$$n = 80, \qquad \Sigma x = 8376, \qquad \Sigma x^2 = 877\,687$$

(i) Calculate unbiased estimates for the population mean and standard deviation.

(ii) State suitable null and alternative hypotheses for a test to see whether the machine appears to be set correctly.

(iii) Carry out the test, at the 10% significance level, stating your conclusions carefully.

11 A magazine conducted a survey about the sleeping time of adults. A random sample of 12 adults was chosen from the adults travelling to work on a train.

(i) Give a reason why this is an unsatisfactory sample for the purposes of the survey.

(ii) State a population for which this sample would be satisfactory.

A satisfactory sample of 12 adults gave numbers of hours of sleep as shown below.

4.6 6.8 5.2 6.2 5.7 7.1 6.3 5.6 7.0 5.8 6.5 7.2

(iii) Calculate unbiased estimates of the mean and variance of the sleeping times of adults.

Cambridge International AS & A Level Mathematics 9709 Paper 7 Q1 June 2008

12 A study of a large sample of books by a particular author shows that the number of words per sentence can be modelled by a normal distribution with mean 21.2 and standard deviation 7.3. A researcher claims to have discovered a previously unknown book by this author. The mean length of 90 sentences chosen at random in this book is found to be 19.4 words.

(i) Assuming the population standard deviation of sentence lengths in this book is also 7.3, test at the 5% level of significance whether the mean sentence length is the same as the author's. State your null and alternative hypotheses.

(ii) State in words relating to the context of the test what is meant by a Type I error and state the probability of a Type I error in the test in part (i).

Cambridge International AS & A Level Mathematics
9709 Paper 7 Q4 June 2005

13 The number of cars caught speeding on a certain length of motorway is 7.2 per day, on average. Speed cameras are introduced and the results shown in the following table are those from a random selection of 40 days after this.

Number of cars caught speeding	4	5	6	7	8	9	10
Number of days	5	7	8	10	5	2	3

(i) Calculate unbiased estimates of the population mean and variance of the number of cars per day caught speeding after the speed cameras were introduced.

(ii) Taking the null hypothesis H_0 to be $\mu = 7.2$, test at the 5% level whether there is evidence that the introduction of speed cameras has resulted in a reduction in the number of cars caught speeding.

(iii) State what is meant by Type I error in words relating to the context of the test in part (ii). Without further calculation, illustrate on a suitable diagram the region representing the probability of this Type I error.

Cambridge International AS & A Level Mathematics
9709 Paper 7 Q7 June 2006

14 A machine has produced nails over a long period of time, where the length in millimetres was distributed as N(22.0, 0.19). It is believed that recently the mean length has changed. To test this belief a random sample of 8 nails is taken and the mean length is found to be 21.7 mm. Carry out a hypothesis test at the 5% significance level to test whether the population mean has changed, assuming that the variance remains the same.

Cambridge International AS & A Level Mathematics
9709 Paper 7 Q3 June 2007

15 In summer the growth rate of grass in a lawn has a normal distribution with mean 3.2 cm per week and standard deviation 1.4 cm per week. A new type of grass is introduced which the manufacturer claims has a slower growth rate. A hypothesis test of this claim at the 5% significance level was carried out using a random sample of 10 lawns that had the new grass. It may be assumed that the growth rate of the new grass has a normal distribution with standard deviation 1.4 cm per week.

(i) Find the rejection region for the test.

(ii) The probability of making a Type II error when the actual value of the mean growth rate of the new grass is m cm per week is less than 0.5. Use your answer to part (i) to write down an inequality for m.

Cambridge International AS & A Level Mathematics 9709 Paper 7 Q2 November 2007

16 Sami claims that he can read minds. He asks each of 50 people to choose one of the 5 letters A, B, C, D or E. He then tells each person which letter he believes they have chosen. He gets 13 correct. Sami says 'This shows that I can read minds, because 13 is more than I would have got right if I were just guessing.'

(i) State null and alternative hypotheses for a test of Sami's claim.

(ii) Test at the 10% significance level whether Sami's claim is justified.

Cambridge International AS & A Level Mathematics 9709 Paper 71 Q2 June 2015

4.2 The Central Limit Theorem

Organic Veg World

The perfect apple grower

Fruit buyer Tom Sisulu writes:

Fruit grower, Rose Ncune, believes that, after years of trials, she has developed trees that will produce the perfect supermarket apple. 'There are two requirements,' Rose told me. 'The average weight of an apple should be 100 grams and they should all be nearly the same size. I have measured hundreds of mine and the standard deviation is a mere 5 grams.'

Rose invited me to take any ten apples off the shelf and weigh them for myself. It was quite uncanny; they were all so close to the magic 100 grams: 98, 107, 105, 98, 100, 99, 104, 93, 105, 103.

Rose is calling her apple the 'Cape Pippin'.

What can you conclude from the weights of Tom's sample of ten apples?

Before going any further, it is appropriate to question whether his sample was random. Rose invited Tom to 'take any ten apples off the shelf'. That is not necessarily the same as taking any ten off the tree. The apples on the shelf could all have been specially selected to impress him. So what follows is based on the assumption that Rose has been honest and the ten apples really do constitute a random sample.

The sample mean is

$$\bar{x} = \frac{98 + 107 + 105 + 98 + 100 + 99 + 104 + 93 + 105 + 103}{10} = 101.2$$

> › What does that tell you about the population mean, μ?

To estimate how far the value of μ is from 101.2, you need to know something about the spread of the data; the usual measure is the standard deviation, σ. In the blog for Organic Veg World you are told that $\sigma = 5$.

The result that if repeated samples of size n are drawn from a population with a normal distribution with mean μ and standard deviation σ, the distribution of the sample means is also normal, its mean is μ and its standard deviation is $\frac{\sigma}{\sqrt{n}}$ is proved in Appendix 6 at www.hoddereducation.com/cambridgeextras.

This is actually a special case of a more general result called the **Central Limit Theorem**. The Central Limit Theorem covers the case where samples are drawn from a population that is not necessarily normal.

» The Central Limit Theorem states that for samples of size n drawn from **any** distribution with mean μ and finite variance σ^2, the distribution of the sample means is approximately $N\left(\mu, \frac{\sigma^2}{n}\right)$ for sufficiently large n.

This theorem is fundamental to much of statistics and so it is worth pausing to make sure you understand just what it is saying.

It deals with the distribution of sample means. This is called the sampling distribution (or more correctly the sampling distribution of the means). There are three aspects to it.

1. The mean of the sample means is μ, the population mean of the original distribution. That is not a particularly surprising result but it is extremely important.
2. The standard deviation of the sample means is $\frac{\sigma}{\sqrt{n}}$. This is often called the **standard error of the mean**.

   Within a sample you would expect some values above the population mean, others below it, so that overall the deviations would tend to cancel each other out, and the larger the sample the more this would be the case. Consequently the standard deviation of the sample means is smaller than that of individual items, by a factor of $\sqrt{n}$.
3. The distribution of sample means is approximately normal.

This last point is the most surprising part of the theorem. Even if the underlying parent distribution is not normal, the distribution of the means of samples of a particular size drawn from it is approximately normal. The larger the sample size, n, the closer this distribution is to the normal. For any given value of n the sampling distribution will be closest to normal where the parent distribution is not unlike the normal.

In many cases the value of n does not need to be particularly large. For most parent distributions you can rely on the distribution of sample means being normal if n is about 20 or 25 (or more).

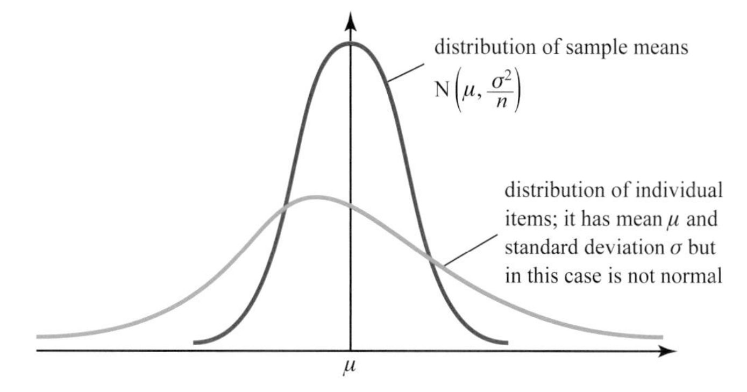

▲ **Figure 4.8**

4.3 Confidence intervals

Returning to the figures on the Cape Pippin apples, you would estimate the population mean to be the same as the sample mean, namely 101.2.

You can express this by saying that you estimate μ to lie within a range of values, an interval, centred on 101.2

$$101.2 - \text{a bit} < \mu < 101.2 + \text{a bit}.$$

Such an interval is called a **confidence interval**.

Imagine you take a large number of samples and use a formula to work out the interval for each of them. If you catch the true population mean in 90% of your intervals, the confidence interval is called a 90% confidence interval. Other percentages are also used and the confidence intervals are named accordingly. The width of the interval is clearly twice the 'bit'.

Finding a confidence interval involves a very simple calculation, but the reasoning behind it is somewhat subtle and requires clear thinking. It is explained in the next section, but you may prefer to make your first reading of it a light one. You should, however, come back to it at some point, otherwise you will not really understand the meaning of confidence intervals.

The theory of confidence intervals

To understand confidence intervals you need to look not at the particular sample whose mean you have just found, but at the parent population from which it was drawn. For the data on the Cape Pippin apples this does not look very promising. All you know about it is its standard deviation σ (in this case 5). You do not know its mean, μ, which you are trying to estimate, or even its shape.

It is now that the strength of the Central Limit Theorem becomes apparent. This states that the distribution of the means of samples of size n drawn from this population is approximately normal with mean μ and standard deviation $\frac{\sigma}{\sqrt{n}}$.

In Figure 4.9 the central 90% region has been shaded leaving the two 5% tails, corresponding to z values of ± 1.645, unshaded. So if you take a large number of samples, all of size n, and work out the sample mean $\bar{x}$ for each one, you would expect that in 90% of cases the value of $\bar{x}$ would lie in the shaded region between A and B.

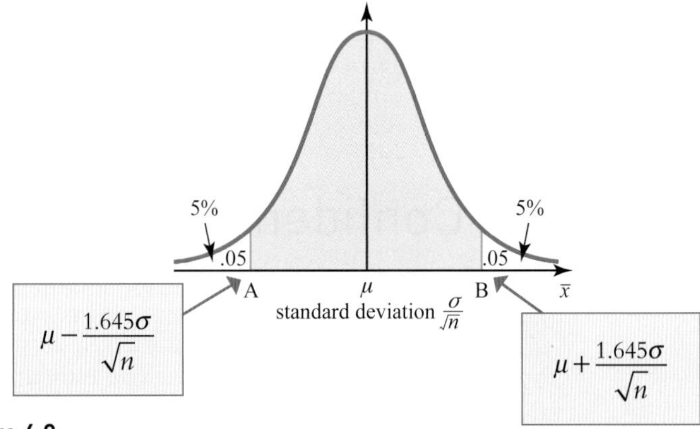

▲ **Figure 4.9**

For such a value of $\bar{x}$ to be in the shaded region

it must be to the right of A: $\bar{x} > \mu - 1.645\frac{\sigma}{\sqrt{n}}$ ①

it must be to the left of B: $\bar{x} < \mu + 1.645\frac{\sigma}{\sqrt{n}}$ ②

Rearranging these two inequalities:

① $\bar{x} + 1.645\frac{\sigma}{\sqrt{n}} > \mu$ or $\mu < \bar{x} + 1.645\frac{\sigma}{\sqrt{n}}$

② $\bar{x} - 1.645\frac{\sigma}{\sqrt{n}} < \mu$

Putting them together gives the result that in 90% of cases

$$\bar{x} - 1.645\frac{\sigma}{\sqrt{n}} < \mu < \bar{x} + 1.645\frac{\sigma}{\sqrt{n}}$$

and this is the 90% confidence interval for μ.

The numbers corresponding to the points A and B are called the 90% **confidence limits** and 90% is the **confidence level**. If you want a different confidence level, you use a different z value from 1.645.

This number is often denoted by k; commonly used values are:

Confidence level	k
90%	1.645
95%	1.960
99%	2.576

and the confidence interval is given by

$$\bar{x} - k\frac{\sigma}{\sqrt{n}} \quad \text{to} \quad \bar{x} + k\frac{\sigma}{\sqrt{n}}.$$

Note

Notice that this is a two-sided symmetrical confidence interval for the mean, μ. Confidence intervals do not need to be symmetrical and can be one-sided. The term 'confidence interval' is a general one, applying not just to the mean but to other population parameters, like variance and skewness, as well. All these cases, however, are outside the scope of this book.

The P% confidence interval for the mean is an interval constructed from sample data in such a way that P% of such intervals will include the true population mean. Figure 4.10 shows a number of confidence intervals constructed from different samples, one of which fails to catch the population mean.

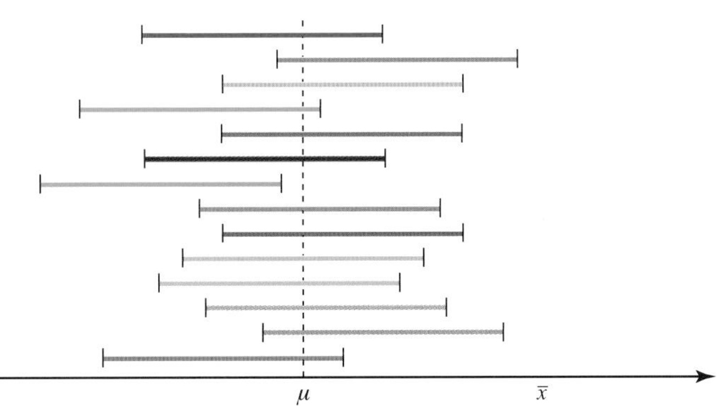

▲ **Figure 4.10**

In the case of the data on the Cape Pippin apples,

$$\bar{x} = 101.2, \qquad \sigma = 5, \qquad n = 10$$

and so the 90% confidence interval is

$$101.2 - 1.645 \times \frac{5}{\sqrt{10}} \quad \text{to} \quad 101.2 + 1.645 \times \frac{5}{\sqrt{10}}$$

$$98.6 \quad \text{to} \quad 103.8.$$

Known and estimated standard deviation

Notice that you can only use this procedure if you already know the value of the standard deviation of the parent population, σ. In this example, Rose Ncune said that she knew, from hundreds of measurements of her apples, that its value is 5.

It is more often the situation that you do not know the population standard deviation or variance and have to estimate it from your sample data. If that is the case, the procedure is different in that you use the *t*-distribution rather than the normal, provided that the parent population is normally distributed, and this results in different values of *k*. However, the use of the *t*-distribution is beyond the scope of this book.

However, if the sample is large, for example over 50, confidence intervals worked out using the normal distribution will be reasonably accurate even though the standard deviation used is an estimate from the sample. So it is quite acceptable to use the normal distribution for large samples whether the standard deviation is known or not.

EXPERIMENTS

These experiments are designed to help you understand confidence intervals, rather than to teach you anything new about dice.
When a single die is thrown, the possible outcomes, 1, 2, 3, 4, 5, 6, are all equally likely with probability $\frac{1}{6}$. Consequently the expectation or mean score from throwing a die is

$$\mu = 1 \times \tfrac{1}{6} + 2 \times \tfrac{1}{6} + \ldots + 6 \times \tfrac{1}{6} = 3.5.$$

Similarly the standard deviation is

$$\sigma = \sqrt{(1^2 \times \tfrac{1}{6} + 2^2 \times \tfrac{1}{6} + \ldots + 6^2 \times \tfrac{1}{6}) - 3.5^2} = 1.708.$$

Imagine that you know σ but don't know μ and wish to construct a 90% confidence interval for it.

Converging confidence intervals

Start by throwing a die once. Suppose you get a 5. You have a sample of size 1, namely {5}, which you could use to work out a sort of 90% confidence interval (but see the warning below).
This confidence interval is given by

$$5 - 1.645 \times \frac{1.708}{\sqrt{1}} \quad \text{to} \quad 5 + 1.645 \times \frac{1.708}{\sqrt{1}}$$

$$2.19 \quad \text{to} \quad 7.81.$$

So far the procedure is not valid. The sample is small and the underlying distribution is not normal. However, things will get better. The more times you throw the die, the larger the sample size and so the more justifiable the procedure.

Now throw the die again. Suppose this time you get a 3. You now have a sample of size 2, namely {5, 3}, with mean 4, and can work out another confidence interval.

The confidence interval is given by

$$4 - 1.645 \times \frac{1.708}{\sqrt{2}} \quad \text{to} \quad 4 + 1.645 \times \frac{1.708}{\sqrt{2}}$$

$$2.79 \quad \text{to} \quad 5.21.$$

Now throw the die again and find a third confidence interval, and a fourth, fifth and so on. You should find them converging on the population mean of 3.5; but it may take some time to get close, particularly if you start with, say, two 6s. This demonstrates that, the larger the sample you take, the narrower the range of values within the confidence interval.

Catching the population mean

Organise a group of friends to throw five dice (or one die five times), and do this 100 times. Each of these gives a sample of size 5 and so you can use it to work out a 90% confidence interval for μ.

You know that the real value of μ is 3.5 and it should be that this is caught within 90% of 90% confidence intervals.

?

› Out of your 100 confidence intervals, how many actually enclose 3.5?

4.4 How large a sample do you need?

You are now in a position to start to answer the question of how large a sample needs to be. The answer, as you will see in Example 4.4, depends on the precision you require, and the confidence level you are prepared to accept.

Example 4.4

A trading standards officer is investigating complaints that a coal merchant is giving short measure. Each sack should contain 25 kg but some variation will inevitably occur because of the size of the lumps of coal; the officer knows from experience that the standard deviation should be 1.5 kg.

The officer plans to take, secretly, a random sample of n sacks, find the total weight of the coal inside them and thereby estimate the mean weight of the coal per sack. He wants to present this figure correct to the nearest kilogram with 95% confidence. What value of n should he choose?

→

Solution

The 95% confidence interval for the mean is given by

$$\bar{x} - 1.96\frac{\sigma}{\sqrt{n}} \quad \text{to} \quad \bar{x} + 1.96\frac{\sigma}{\sqrt{n}}$$

and so, since $\sigma = 1.5$, the inspector's requirement is that

$$\frac{1.96 \times 1.5}{\sqrt{n}} \leqslant 0.5$$

$$\Rightarrow \quad \frac{1.96 \times 1.5}{0.5} \leqslant \sqrt{n}$$

$$\Rightarrow \quad n \geqslant 34.57.$$

So the inspector needs to take 35 sacks.

Large samples

Given that the width of a confidence interval decreases with sample size, why is it not standard practice to take very large samples?

The answer is that the cost and time involved have to be balanced against the quality of information produced. Because the width of a confidence interval depends on $\frac{1}{\sqrt{n}}$ and not on $\frac{1}{n}$, increasing the sample size does not produce a proportional reduction in the width of the interval. You have, for example, to increase the sample size by a factor of 4 to halve the width of the interval. In the previous example the inspector had to weigh 35 sacks of coal to achieve a class interval of $2 \times 0.5 = 1$ kg with 95% confidence. That is already quite a daunting task; does the benefit from reducing the interval to 0.5 kg justify the time, cost and trouble involved in weighing another 105 sacks?

4.5 Confidence intervals for a proportion

In this chapter you have seen how to calculate confidence intervals for the population mean. Confidence intervals can also be found for other population parameters, like the variance or, in a binomial situation, the proportion of the population with a particular characteristic.

In Example 4.5 a confidence interval for a population proportion is found. The method assumes that the sample taken is large, and so the normal approximation to the binomial distribution may be used.

In an experiment where a (large) sample of size n has been taken and m items are found to have the characteristic under investigation, the population proportion is estimated to be $p = \frac{m}{n}$.

For all samples of size n, the following estimates may then be made.

- Mean number of occurrences per sample = np
- Variance of number of occurrences per sample = $npq = np(1-p)$
- Variance of estimated proportion $= \dfrac{np(1-p)}{n^2} = \dfrac{p(1-p)}{n}$
- Standard deviation of estimated proportion $= \sqrt{\dfrac{p(1-p)}{n}}$

So the confidence interval for the proportion is given by

$$p - k\sqrt{\frac{p(1-p)}{n}} < \text{population proportion} < p + k\sqrt{\frac{p(1-p)}{n}},$$

where the values of k are taken from the normal distribution: 1.96 for a 95% (two-sided) confidence interval, 1.645 for a 90% interval, etc.

Example 4.5

A certain type of moth is found in two colours, brown and white. In an experiment, 100 moths from a particular region are captured. Thirty of them are found to be brown, the remainder white.

Calculate the 95% confidence interval for the population proportion of brown moths.

Solution

Estimated population proportion of brown moths, $p = \frac{30}{100} = 0.3$.

So the 95% confidence interval is given by

$$0.3 - 1.96\sqrt{\frac{0.3(1-0.3)}{100}} < \text{population proportion} < 0.3 + 1.96\sqrt{\frac{0.3(1-0.3)}{100}}$$

giving

$$0.210 < \text{population proportion} < 0.390 \text{ (to 3 s.f.).}$$

Exercise 4B

1 A biologist studying a colony of beetles selects and weighs a random sample of 20 adult males. She knows that, because of natural variability, the weights of such beetles are normally distributed with standard deviation 0.2 g. Their weights, in grams, are as follows.

5.2	5.4	4.9	5.0	4.8	5.7	5.2	5.2	5.4	5.1
5.6	5.0	5.2	5.1	5.3	5.2	5.1	5.3	5.2	5.2

(i) Find the mean weight of the beetles in this sample.

(ii) Find 95% confidence limits for the mean weight of such beetles.

M **2** An aptitude test for deep-sea divers has been designed to produce scores that are approximately normally distributed on a scale from 0 to 100 with standard deviation 25. The scores from a random sample of people taking the test were as follows.

23 35 89 35 12 45 60 78 34 66

(i) Find the mean score of the people in this sample.

(ii) Construct a 90% confidence interval for the mean score of people taking the test.

(iii) Construct a 99% confidence interval for the mean score of people taking the test. Compare this confidence interval with the 90% confidence interval.

M **3** A manufacturer of women's clothing wants to know the mean height of the women in a town (in order to plan what proportion of garments should be of each size). She knows that the standard deviation of their heights is 5 cm. She selects a random sample of 50 women from the town and finds their mean height to be 165.2 cm.

(i) Use the available information to estimate the proportion of women in the town who were

(a) over 170 cm tall (b) less than 155 cm tall.

(ii) Construct a 95% confidence interval for the mean height of women in the town.

(iii) Another manufacturer in the same town wants to know the mean height of women in the town to within 0.5 cm with 95% confidence. What is the minimum sample size that would ensure this?

M **4** An examination question, marked out of 10, is answered by a very large number of candidates. A random sample of 400 scripts is taken and the marks on this question are recorded.

Mark	0	1	2	3	4	5	6	7	8	9	10
Frequency	12	35	11	12	3	20	57	87	20	14	129

(i) Calculate the sample mean and the sample standard deviation.

(ii) Assuming that the population standard deviation has the same value as the sample standard deviation, find 90% confidence limits for the population mean.

CP **5** An archaeologist discovers a short manuscript in an ancient language that he recognises but cannot read. There are 30 words in the manuscript and they contain a total of 198 letters. There are two written versions of the language. In the *early* form of the language the mean word length is 6.2 letters with standard deviation 2.5; in the *late* form words were given prefixes, raising the mean length to 7.6 letters but leaving the standard deviation unaltered. The archaeologist hopes the manuscript will help him to date the site.

(i) Construct a 95% confidence interval for the mean word length of the language used in the short manuscript.

(ii) What advice would you give the archaeologist?

CP **6** A football boot manufacturer did extensive testing on the wear of the front studs of its Supa range. It found that, after 30 hours' use, the wear (i.e. the amount by which the length was reduced) was normally distributed with standard deviation 1.3 mm. However, the mean wear on the studs of the boot on the dominant foot of the player was 4 mm more than on the studs of the other boot and the standard deviation of the difference in wear between a pair of boots was 1.838 mm.

The coach of a football team accepted the claim for the standard deviation but was suspicious of the claim about the mean difference. He chose ten of his squad at random. He fitted them with new boots and measured the wear after 30 hours of use with the following results.

Player	1	2	3	4	5	6	7	8	9	10
Dominant foot	6.5	8.3	4.5	6.7	9.2	5.3	7.6	8.1	9.0	8.4
Other foot	4.2	4.6	2.3	3.8	7.0	4.7	1.4	3.8	8.4	5.7

(i) Calculate 95% confidence limits for the mean difference in wear based on the sample data.

(ii) Use these limits to explain whether or not you consider the coach's suspicions were justified.

CP **7** A school decided to introduce a new activity programme for its new students to try to improve the fitness of the students. In order to see whether the programme was effective, several tests were done. For one of these, the students were timed on a run of 1 kilometre in their first week in the school and again ten weeks later. A random sample of 100 of the students did both runs. The differences of their mean times, subtracting the time of the second run from that of the first, were calculated. The mean and standard deviation were found to be 0.75 minutes and 1.62 minutes respectively.

Calculate a 90% confidence interval for the population mean difference. You may assume that the differences are distributed normally. What assumption have you made in finding this confidence interval?

The organiser of the programme considers that it should lead to an improvement of at least half a minute in the average times. Explain whether or not this aim has been achieved.

CP

8 In an experiment to see if reaction times were affected by whether or not individuals are hungry, 2000 randomly chosen soldiers were tested before and after they had eaten a substantial lunch. The test used was to drop a metre rule, which was held vertically so that its lower end was level with the thumb and first finger of each person, and to measure how far the rule fell before it was caught. For each person, the difference, d, of the distance measured after lunch minus the distance measured before lunch was found. From these it was calculated that $\Sigma d = 1626$ and $\Sigma d^2 = 258\,632$.

Use these data to provide a 98% confidence interval for the population mean difference, stating any assumptions you have made.

What does your confidence interval suggest about reaction times before and after a meal?

9 The weights in grams of oranges grown in a certain area are normally distributed with mean μ and standard deviation σ. A random sample of 50 of these oranges was taken, and a 97% confidence interval for μ based on this sample was (222.1, 232.1).

(i) Calculate unbiased estimates of μ and σ^2.

(ii) Estimate the sample size that would be required in order for a 97% confidence interval for μ to have width 8.

Cambridge International AS & A Level Mathematics 9709 Paper 71 Q2 June 2009

10 (i) Give a reason why, in carrying out a statistical investigation, a sample rather than a complete population may be used.

(ii) Rose wishes to investigate whether men in her town have a different life-span from the national average of 71.2 years. She looks at government records for her town and takes a random sample of the ages of 110 men who have died recently. Their mean age in years was 69.3 and the unbiased estimate of the population variance was 65.61.

(a) Calculate a 90% confidence interval for the population mean and explain what you understand by this confidence interval.

(b) State with a reason what conclusion about the life-span of men in her town Rose could draw from this confidence interval.

Cambridge International AS & A Level Mathematics 9709 Paper 7 Q4 November 2005

11 Diameters of golf balls are known to be normally distributed with mean μ cm and standard deviation σ cm. A random sample of 130 golf balls was taken and the diameters, x cm, were measured. The results are summarised by $\Sigma x = 555.1$ and $\Sigma x^2 = 2371.30$.

(i) Calculate unbiased estimates of μ and σ^2.

(ii) Calculate a 97% confidence interval for μ.

(iii) 300 random samples of 130 golf balls are taken and a 97% confidence interval is calculated for each sample. How many of these intervals would you expect **not** to contain μ?

Cambridge International AS & A Level Mathematics
9709 Paper 7 Q4 November 2008

12 A random sample of n people were questioned about their internet use. 87 of them had a high-speed internet connection. A confidence interval for the population proportion having a high-speed internet connection is $0.1129 < p < 0.1771$.

(i) Write down the midpoint of this confidence interval and hence find the value of n.

(ii) This interval is an α% confidence interval. Find α.

Cambridge International AS & A Level Mathematics
9709 Paper 71 Q2 June 2010

13 (i) Explain what is meant by the term 'random sample'.

In a random sample of 350 food shops it was found that 130 of them had Special Offers.

(ii) Calculate an approximate 95% confidence interval for the proportion of all food shops with Special Offers.

(iii) Estimate the size of a random sample required for an approximate 95% confidence interval for this proportion to have a width of 0.04.

Cambridge International AS & A Level Mathematics
9709 Paper 7 Q3 November 2007

14 A survey of a random sample of n people found that 61 of them read *The Reporter* newspaper. A symmetric confidence interval for the true population proportion, p, who read *The Reporter* is $0.1993 < p < 0.2887$.

(i) Find the midpoint of this confidence interval and use this to find the value of n.

(ii) Find the confidence level of this confidence interval.

Cambridge International AS & A Level Mathematics
9709 Paper 7 Q3 June 2005

15 The masses, m grams, of a random sample of 80 strawberries of a certain type were measured and summarised as follows.

$$n = 80 \qquad \Sigma m = 4200 \qquad \Sigma m^2 = 229\,000$$

(i) Find unbiased estimates of the population mean and variance.

(ii) Calculate a 98% confidence interval for the population mean.

50 random samples of size 80 were taken and a 98% confidence interval for the population mean, μ, was found from each sample.

(iii) Find the number of these 50 confidence intervals that would be expected to include the true value of μ.

Cambridge International AS & A Level Mathematics
9709 Paper 71 Q5 June 2015

KEY POINTS

1 Distribution of sample means

- For samples of size n drawn from a normal distribution with mean μ and finite variance σ^2, the distribution of sample means is normal with mean μ and variance $\frac{\sigma^2}{n}$, i.e. $\bar{x} \sim \mathrm{N}\left(\mu, \frac{\sigma^2}{n}\right)$.
- The standard error of the mean (i.e. the standard deviation of the sample means) is given by $\frac{\sigma}{\sqrt{n}}$.

2 Hypothesis testing

- Sample data may be used to carry out a hypothesis test on the null hypothesis that the population mean has some particular value, μ_0, i.e. H_0: $\mu = \mu_0$.
- The test statistic is $z = \frac{\bar{x} - \mu_0}{\frac{\sigma}{\sqrt{n}}}$ and the normal distribution is used.

For situations where the population mean, μ, is unknown but the population variance, σ^2 (or standard deviation, σ), is known:

3 The Central Limit Theorem

For samples of size n drawn from any distribution with mean μ and finite variance σ^2, the distribution of the sample means is approximately

$\mathrm{N}\left(\mu, \frac{\sigma^2}{n}\right)$ for sufficiently large n.

4 The standard error of the mean

The standard error of the mean (i.e. the standard deviation of the sample means) is given by $\frac{\sigma}{\sqrt{n}}$.

5 Confidence intervals

Two-sided confidence intervals for μ are given by

$$\overline{x} - k\frac{\sigma}{\sqrt{n}} \quad \text{to} \quad \overline{x} + k\frac{\sigma}{\sqrt{n}}.$$

The value of k for any confidence level can be found using normal distribution tables.

Confidence level	k
90%	1.645
95%	1.960
99%	2.576

LEARNING OUTCOMES

Now that you have finished this chapter, you should be able to

- formulate hypotheses and carry out a hypothesis test for the population mean of a normal distribution
- use the normal distribution as an approximation to test for the mean, and associated proportion, of a binomial distribution, applying continuity corrections where appropriate
- understand that the sample mean is a random variable
- know the distribution of the mean of samples of size n form a normal distribution
- use the normal distribution as an approximation to the binomial distribution
- carry out a hypothesis test for a single mean using the normal distribution, using either a p-value or a critical region:
  - where the population variance is known
  - where the population variance is unknown but the sample size is large
- calculate the probabilities of making Type I and Type II errors in specific situations involving tests based on a normal distribution
- know that $E(\overline{X}) = \mu$ and $\text{Var}(\overline{X}) = \frac{\sigma^2}{n}$
- use the Central Limit Theorem
- calculate unbiased estimates of the population mean and variance from a sample
- find a confidence interval for a population mean:
  - where the population variance is known
  - where the population variance is unknown but the sample size is large
- find a confidence interval for a population proportion.

5 The Poisson distribution

> If something can go wrong, sooner or later it will go wrong.
> *Murphy's Law*

Electrics Express Online

Since our 'next day delivery guarantee' went live, the number of orders has increased dramatically. We are now one of the most popular websites for mail-order electrical goods. We would like to reassure our customers that we have taken on more staff to cope with the increased demand for our products. It is impossible to predict the level of demand; however, we do know that we are receiving an average of 150 orders per hour!

The appearance of this update on their website prompted a statistician to contact Electrics Express Online. She offered to analyse the data and see what suggestions she could come up with.

For her detailed investigation, she considered the distribution of the number of orders per minute. For a random sample of 1000 single-minute intervals during the last month, she collected the following data.

Number of orders per minute	0	1	2	3	4	5	6	7	>7
Frequency	70	215	265	205	125	75	30	10	5

Summary statistics for this frequency distribution are as follows.

$$n = 1000, \quad \sum xf = 2525 \quad \text{and} \quad \sum x^2 f = 8885$$

$$\Rightarrow \quad \bar{x} = 2.525 \quad \text{and} \quad sd = 1.58 \text{ (to 3 s.f.)}$$

She also noted that:

- orders made on the website appear at random and independently of each other
- the average number of orders per minute is about 2.5, which is equivalent to 150 per hour.

She suggested that the appropriate probability distribution to model the number of orders was the Poisson distribution.

The particular Poisson distribution, with an average number of 2.5 orders per minute, is defined as an **infinite** discrete random variable given by

$$P(X = r) = e^{-2.5} \times \frac{2.5^r}{r!} \quad \text{for} \quad r = 0, 1, 2, 3, 4, \ldots$$

where

- X represents the random variable 'number of orders per minute'
- e is the mathematical constant 2.718 281 828 459...
- $e^{-2.5}$ can be found from your calculator as 0.082 (to 3 d.p.)
- $r!$ means r **factorial**, for example $5! = 5 \times 4 \times 3 \times 2 \times 1 = 120$.

Values of the corresponding probability distribution may be tabulated using the formula, together with the expected frequencies this would generate. For example,

$$\begin{aligned} P(X = 4) &= e^{-2.5} \times \frac{2.5^4}{4!} \\ &= 0.133\,60 \ldots \\ &= 0.134 \text{ (to 3 s.f.)} \end{aligned}$$

Number of orders per minute (r)	0	1	2	3	4	5	6	7	>7
Observed frequency	70	215	265	205	125	75	30	10	5
P($X = r$)	0.082	0.205	0.257	0.214	0.134	0.067	0.028	0.010	0.003
Expected frequency	82	205	257	214	134	67	28	10	3

The closeness of the observed and expected frequencies (see Figure 5.1) implies that the Poisson distribution is indeed a suitable model in this instance.

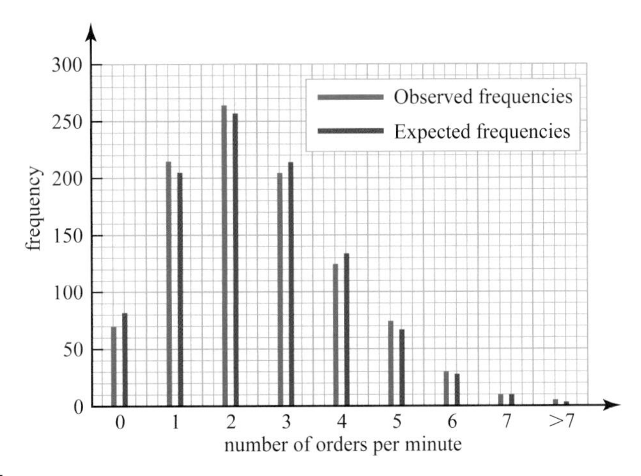

▲ **Figure 5.1**

Note also that the sample mean, $\bar{x} = 2.525$, is very close to the sample variance, $s^2 = 2.509$ (to 4 s.f.). You will see later that, for a Poisson distribution, the expectation and variance are the same. So the closeness of these two summary statistics provides further evidence that the Poisson distribution is a suitable model.

5.1 The Poisson distribution

A discrete random variable may be modelled by a Poisson distribution provided:

- events occur at random and independently of each other, in a given interval of time or space
- the average number of events in the given interval, λ, is uniform and finite.

Let X represent the number of occurrences in a given interval, then

$$P(X = r) = e^{-\lambda} \times \frac{\lambda^r}{r!} \quad \text{for} \quad r = 0, 1, 2, 3, 4, \ldots$$

Like the discrete random variables you met in *Probability & Statistics 1*, the Poisson distribution may be illustrated by a vertical line chart. The shape of the Poisson distribution depends on the value of the parameter λ (pronounced 'lambda'). The letter μ (pronounced 'mu') is also commonly used to represent the Poisson parameter. If λ is small the distribution has positive skew, but as λ increases the distribution becomes progressively more symmetrical. Three typical Poisson distributions are illustrated in Figure 5.2.

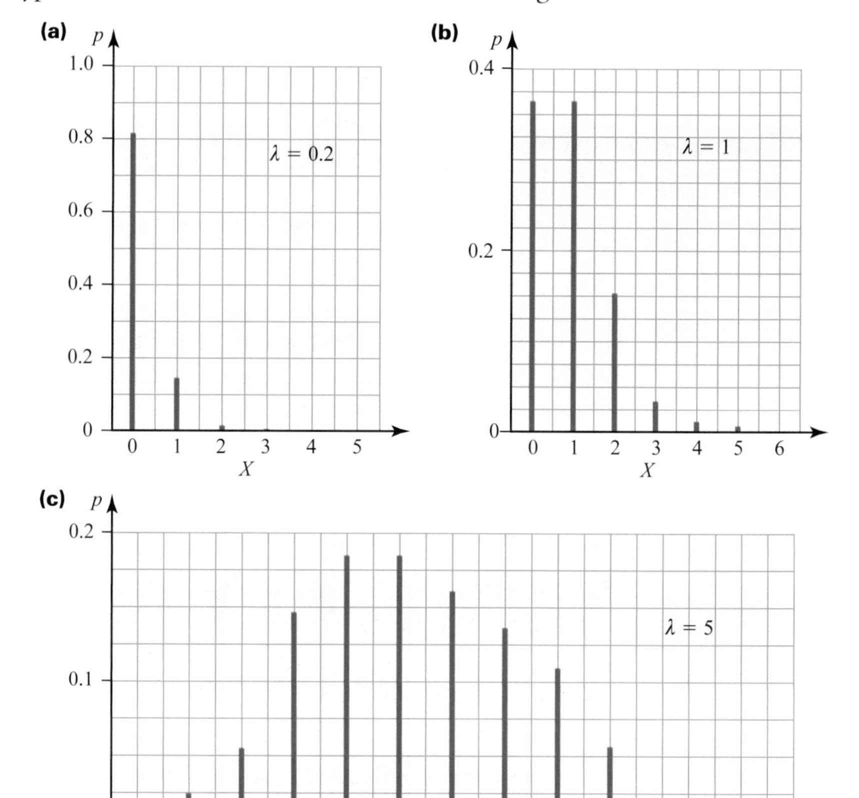

▲ **Figure 5.2** The shape of the Poisson distribution for (a) $\lambda = 0.2$ (b) $\lambda = 1$ (c) $\lambda = 5$

There are many situations in which events happen singly and the average number of occurrences per given interval of time or space is uniform and is known or can be easily found. Such events might include: the number of goals scored by a team in a football match, the number of telephone calls received per minute at an exchange, the number of accidents in a factory per week, the number of particles emitted in a minute by a radioactive substance whose half-life is relatively long, the number of typing errors per page in a document, the number of flaws per metre in a roll of cloth or the number of micro-organisms in 1 millilitre of pond water.

Example 5.1

The number of defects in a wire cable can be modelled by the Poisson distribution with a uniform rate of 1.5 defects per kilometre.

Find the probability that

(i) a single kilometre of wire will have exactly three defects

(ii) a single kilometre of wire will have at least five defects.

Solution

Let X represent the number of defects per kilometre, then

$$P(X = r) = e^{-1.5} \times \frac{1.5^r}{r!} \quad \text{for} \quad r = 0, 1, 2, 3, 4, \ldots.$$

(i) $$P(X = 3) = e^{-1.5} \times \frac{1.5^3}{3!}$$
$$= 0.125\,510\ldots$$
$$= 0.126 \text{ (to 3 s.f.)}$$

(ii) $$P(X \geqslant 5) = 1 - [P(X = 0) + P(X = 1) + P(X = 2) + P(X = 3) + P(X = 4)]$$
$$= 1 - \left[e^{-1.5} \times \frac{1.5^0}{0!} + e^{-1.5} \times \frac{1.5^1}{1!} + e^{-1.5} \times \frac{1.5^2}{2!} + e^{-1.5} \times \frac{1.5^3}{3!} + e^{-1.5} \times \frac{1.5^4}{4!}\right]$$
$$= 1 - [0.223\,130\ldots + 0.334\,695\ldots + 0.251\,021\ldots + 0.125\,510\ldots + 0.047\,066\ldots]$$
$$= 0.0186 \text{ (to 3 s.f.)}$$

Calculating Poisson distribution probabilities

In Example 5.1, about the defects in a wire cable, you had to work out $P(X \geqslant 5)$. To do this you used $P(X \geqslant 5) = 1 - P(X \leqslant 4)$, which saved you having to work out all the probabilities for five or more occurrences and adding them together. Such calculations can take a long time even though the terms eventually get smaller and smaller, so that after some time you will have gone far enough for the accuracy you require and may stop.

However, Example 5.1 did involve working out and summing five probabilities and so was quite time consuming. A way of cutting down on the amount of work, and so on the time you take, follows.

Recurrence relations

Recurrence relations allow you to use the term you have obtained to work out the next one. For the Poisson distribution with parameter λ,

$P(X=0) = e^{-\lambda}$ You must use your calculator to find this term.

$P(X=1) = e^{-\lambda} \times \lambda = \lambda P(X=0)$ Multiply the previous term by λ.

$P(X=2) = e^{-\lambda} \times \frac{\lambda^2}{2!} = \frac{\lambda}{2} P(X=1)$ Multiply the previous term by $\frac{\lambda}{2}$.

$P(X=3) = e^{-\lambda} \times \frac{\lambda^3}{3!} = \frac{\lambda}{3} P(X=2)$ Multiply the previous term by $\frac{\lambda}{3}$.

$P(X=4) = e^{-\lambda} \times \frac{\lambda^4}{4!} = \frac{\lambda}{4} P(X=3)$ Multiply the previous term by $\frac{\lambda}{4}$.

In general, you can find $P(X=r)$ by multiplying your previous probability, $P(X=r-1)$, by $\frac{\lambda}{r}$. You would expect to hold the latest value on your calculator and keep a running total in the memory.

Setting this out on paper with $\lambda = 1.5$ (the figure from Example 5.1) gives these figures.

No. of cases, r	Conversion	$P(X=r)$	Running total, $P(X \leqslant r)$
0		0.223 130...	0.223 130...
1	$\times 1.5$	0.334 695...	0.557 825...
2	$\times \frac{1.5}{2}$	0.251 021...	0.808 846...
3	$\times \frac{1.5}{3}$	0.125 510...	0.934 356...
4	$\times \frac{1.5}{4}$	0.047 066...	0.981 422...

Adapting the Poisson distribution for different time intervals

Example 5.2

Jasmit is considering buying a telephone answering machine. He has one for five days' free trial and finds that 22 messages are left on it. Assuming that this is typical of the use it will get if he buys it, find

(i) the mean number of messages per day
(ii) the probability that on one particular day there will be exactly six messages
(iii) the probability that there will be exactly six messages in two days.

Solution

(i) Converting the total for five days to the mean for a single day gives

$$\text{daily mean} = \frac{22}{5} = 4.4 \text{ messages per day}$$

(ii) Calling X the number of messages per day,

$$P(X = 6) = e^{-4.4} \times \frac{4.4^6}{6!}$$
$$= 0.124$$

(iii) The mean for two days is

$$2 \times \frac{22}{5} = 8.8 \text{ messages}$$

So the probability of exactly six messages is

$$e^{-8.8} \times \frac{8.8^6}{6!} = 0.0972$$

5.2 Modelling with a Poisson distribution

In the example about Electrics Express Online, the mean and variance of the number of orders placed per minute on the website were given by $\bar{x} = 2.525$ and $s^2 = 2.51$ (to 3 s.f.). The corresponding Poisson parameter, λ, was then taken to be 2.5.

It can be shown that for any Poisson distribution

$$\text{Mean} = E(X) = \lambda \quad \text{and} \quad \text{Variance} = \text{Var}(X) = \lambda.$$

The notation Po(λ) or Poisson(λ) is used to describe this distribution. Formal derivations of the mean and variance of a Poisson distribution are given in Appendix 4 at www.hoddereducation.com/cambridgeextras.

When modelling data with a Poisson distribution, the closeness of the mean and variance is one indication that the data fit the model well.

When you have collected the data, go through the following steps in order to check whether the data may be modelled by a Poisson distribution.

- Work out the mean and variance and check that they are roughly equal.
- Use the sample mean to work out the Poisson probability distribution and a suitable set of expected frequencies.
- Compare these expected frequencies with your observations.

Exercise 5A

1 If $X \sim \text{Po}(1.75)$, use the Poisson formula to calculate

(i) $P(X = 2)$ (ii) $P(X > 0)$.

2 If $X \sim \text{Po}(3.1)$, use the Poisson formula to calculate

(i) $P(X = 3)$ (ii) $P(X < 2)$ (iii) $P(X \geqslant 2)$.

M **3** The number of wombats that are killed on a particular stretch of road in Australia in any one day can be modelled by a Po(0.42) random variable.

(i) Calculate the probability that exactly two wombats are killed on a given day on this stretch of road.

(ii) Find the probability that exactly four wombats are killed over a five-day period on this stretch of road.

CP **4** A typesetter makes 1500 mistakes in a book of 500 pages. On how many pages would you expect to find (i) 0 (ii) 1 (iii) 2 (iv) 3 or more mistakes? State any assumptions in your workings.

M **5** In a country the mean number of deaths per year from lightning strike is 2.2.

(i) Find the probabilities of 0, 1, 2 and more than 2 deaths from lightning strike in any particular year.

In a neighbouring country, it is found that one year in twenty nobody dies from lightning strike.

(ii) Estimate the mean number of deaths per year in that country from lightning strike.

PS **6** 350 raisins are put into a mixture which is well stirred and made into 100 small buns. Estimate how many of these buns will

(i) be without raisins

(ii) contain five or more raisins.

In a second batch of 100 buns, exactly one has no raisins in it.

(iii) Estimate the total number of raisins in the second mixture.

M **7** A ferry takes cars and small vans on a short journey from an island to the mainland. On a representative sample of weekday mornings, the numbers of vehicles, X, on the 8 am sailing were as follows.

20	24	24	22	23	21	20	22	23	22
21	21	22	21	23	22	20	22	20	24

(i) Show that X does not have a Poisson distribution.

In fact, 20 of the vehicles belong to commuters who use that sailing of the ferry every weekday morning. The random variable Y is the number of vehicles other than those 20 who are using the ferry.

(ii) Investigate whether Y may reasonably be modelled by a Poisson distribution.

The ferry can take 25 vehicles on any journey.

(iii) On what proportion of days would you expect at least one vehicle to be unable to travel on this particular sailing of the ferry because there was no room left and so have to wait for the next one?

8 People arrive randomly and independently at the elevator in a block of flats at an average rate of 4 people every 5 minutes.

(i) Find the probability that exactly two people arrive in a 1-minute period.

(ii) Find the probability that nobody arrives in a 15-second period.

(iii) The probability that at least one person arrives in the next t minutes is 0.9. Find the value of t.

Cambridge International AS & A Level Mathematics 9709 Paper 7 Q6 June 2008

9 A shopkeeper sells electric fans. The demand for fans follows a Poisson distribution with mean 3.2 per week.

(i) Find the probability that the demand is exactly 2 fans in any one week.

(ii) The shopkeeper has 4 fans in his shop at the beginning of a week. Find the probability that this will not be enough to satisfy the demand for fans in that week.

(iii) Given instead that he has n fans in his shop at the beginning of a week, find, by trial and error, the least value of n for which the probability of his not being able to satisfy the demand for fans in that week is less than 0.05.

Cambridge International AS & A Level Mathematics 9709 Paper 7 Q6 November 2005

10 People arrive randomly and independently at a supermarket checkout at an average rate of 2 people every 3 minutes.

(i) Find the probability that exactly 4 people arrive in a 5-minute period.

At another checkout in the same supermarket, people arrive randomly and independently at an average rate of 1 person each minute.

(ii) Find the probability that a total of fewer than 3 people arrive at the two checkouts in a 3-minute period.

Cambridge International AS & A Level Mathematics 9709 Paper 71 Q2 November 2010

5.3 The sum of two or more Poisson distributions

Safer crossing near our school?

A recent traffic survey has revealed that the number of vehicles using the main road outside the school has reached levels where crossing has become a hazard to our students.

The survey, carried out by a group of our students, shows that the volume of traffic has increased so much that our students are almost taking their lives in their hands when crossing the road.

At 3 pm, usually one of the quieter periods of the day, the average number of vehicles passing our school to go into the town is 3.5 per minute and the average number of vehicles heading out of town is 5.7 per minute. A safe crossing is a must!

The town council has told our students that if they can show that there is a greater than 1 in 4 chance of more than 10 vehicles passing per minute, then we should be successful in getting a safer crossing outside the school.

Assuming that the flow of vehicles, into and out of town, can be modelled by independent Poisson distributions, you can model the flow of vehicles in both directions as follows.

Let X represent the number of vehicles travelling into town at 3 pm, then $X \sim \text{Po}(3.5)$.

Let Y represent the number of vehicles travelling out of town at 3 pm, then $Y \sim \text{Po}(5.7)$.

Let T represent the number of vehicles travelling in either direction at 3 pm, then $T = X + Y$.

You can find the probability distribution for T as follows.

$$\begin{aligned}\text{P}(T=0) &= \text{P}(X=0)\times\text{P}(Y=0)\\ &= 0.0302\times 0.0033 = 0.0001\end{aligned}$$

$$\begin{aligned}\text{P}(T=1) &= \text{P}(X=0)\times\text{P}(Y=1)+\text{P}(X=1)\times\text{P}(Y=0)\\ &= 0.0302\times 0.0191 + 0.1057\times 0.0033 = 0.0009\end{aligned}$$

$$\begin{aligned}\text{P}(T=2) &= \text{P}(X=0)\times\text{P}(Y=2)+\text{P}(X=1)\times\text{P}(Y=1)+\text{P}(X=2)\times\text{P}(Y=0)\\ &= 0.0302\times 0.0544 + 0.1057\times 0.0191 + 0.1850\times 0.0033\\ &= 0.0043\end{aligned}$$

and so on.

You can see that this process is very time consuming. Fortunately, you can make life a lot easier by using the fact that if X and Y are two independent

Poisson random variables, with means λ and μ respectively, then if $T = X + Y$ then T is a Poisson random variable with mean $\lambda + \mu$.

$$X \sim \text{Po}(\lambda) \text{ and } Y \sim \text{Po}(\mu) \quad \Rightarrow \quad X + Y \sim \text{Po}(\lambda + \mu)$$

Using $T \sim \text{Po}(9.2)$ gives the required probabilities straight away.

$$\text{P}(T = 0) = \text{e}^{-9.2} = 0.0001$$
$$\text{P}(T = 1) = \text{e}^{-9.2} \times 9.2 = 0.0009$$
$$\text{P}(T = 2) = \text{e}^{-9.2} \times \frac{9.2^2}{2!} = 0.0043$$

and so on.

You can now use the distribution for T to find the probability that the total traffic flow exceeds 10 vehicles per minute.

$$\text{P}(T > 10) = 1 - \text{P}(T \leqslant 10)$$
$$= 1 - 0.6820 = 0.318$$

Since there is a greater than 25% chance of more than 10 vehicles passing per minute, the case for the crossing has been made, based on the Poisson probability models.

Example 5.3

A rare disease causes the death, on average, of 2.0 people per year in Sweden, 0.8 in Norway and 0.5 in Finland. As far as is known the disease strikes at random and cases are independent of one another.

What is the probability of four or more deaths from the disease in these three countries in any year?

Solution

Notice first that:

- P(4 or more deaths) = 1 − P(3 or fewer deaths)
- each of the three distributions fulfils the conditions for it to be modelled by the Poisson distribution.

You can therefore add the three distributions together and treat the result as a single Poisson distribution.

The overall mean is given by	2.0	+	0.8	+	0.5	=	3.3
	Sweden		Norway		Finland		Total

giving an overall distribution of Po(3.3).

The probability of 4 or fewer deaths is then

$$1 - \text{e}^{-3.3} \times \left(1 + 3.3 + \frac{3.3^2}{2!} + \frac{3.3^3}{3!}\right)$$

So the probability of 4 or more deaths is given by

$$1 - 0.580 = 0.420$$

Notes

1. You may only add Poisson distributions in this way if they are independent of each other.
2. The proof of the validity of adding Poisson distributions in this way is given in Appendix 5 at www.hoddereducation.com/cambridgeextras.

Example 5.4

On a lonely Highland road in Scotland, cars are observed passing at the rate of 6 per day and lorries at the rate of 3 per day. On the road is an old cattle grid that will soon need repair. The local works department decides that if the probability of more than 2 vehicles per hour passing is less than 1% then the repairs to the cattle grid can wait until next spring, otherwise it will have to be repaired before the winter.

When will the cattle grid have to be repaired?

Solution

Let C be the number of cars per hour, L be the number of lorries per hour and V be the number of vehicles per hour.

$$V = L + C$$

Assuming that a car or a lorry passing along the road is a random event and the two are independent

$$C \sim \text{Po}(0.25),\ L \sim \text{Po}(0.125)$$

and so $\quad V \sim \text{Po}(0.25 + 0.125)$

$\Rightarrow \quad V \sim \text{Po}(0.375)$

6 cars a day is $\frac{6}{24} = 0.25$ cars in an hour. Similarly, there are $\frac{3}{24} = 0.125$ lorries per hour.

The required probability is

$$\begin{aligned} \text{P}(V > 2) &= 1 - \text{P}(V \leqslant 2) \\ &= 1 - e^{-0.375} \times \left(1 + 0.375 + \frac{0.375^2}{2!}\right) \\ &= 0.006\,65 \end{aligned}$$

This is less than 1% and so the repairs are left until spring.

?

› The modelling of this situation raises a number of questions.

(i) Is it true that a car or lorry passing along the road is a random event, or are some of these regular users, like the lorry collecting the milk from the farms along the road? If, say, three of the cars and one lorry are regular daily users, what effect does this have on the calculation?

(ii) Is it true that every car or lorry travels independently of every other one?

(iii) Are vehicles more likely in some hours than others?

(iv) There are no figures for bicycles or motorcycles or other vehicles. Why might this be so?

Exercise 5B

M

1 A mobile valet service visits an office car park every Monday. It offers exterior-only washes and full valets. Purchases of exterior-only washes and full valets have independent Poisson distributions with means 7.2 and 4.1 respectively.

(i) Find the probabilities that, on any Monday

(a) exactly seven customers purchase exterior-only washes

(b) at least two full valets are purchased

(c) exactly seven customers purchase exterior-only washes and exactly three customers purchase full valets.

(ii) By using the distribution of the total number of customers, find the probability that exactly ten customers use the mobile service on any Monday.

(iii) Given that exactly ten customers use the service on one Monday, find the probability that exactly seven of them purchase exterior-only washes.

M

2 Telephone calls reach a departmental administrator independently and at random, internal ones at a mean rate of two in any five-minute period, and external ones at a mean rate of one in any five-minute period.

(i) Find the probability that in a five-minute period, the administrator receives

(a) exactly three internal calls

(b) at least two external calls

(c) at most five calls in total.

(ii) Given that the administrator receives a total of four calls in a five-minute period, find the probability that exactly two were internal calls.

(iii) Find the probability that in any one-minute interval no calls are received.

3 Two random variables, X and Y, have independent Poisson distributions given by $X \sim \text{Po}(1.4)$ and $Y \sim \text{Po}(3.6)$ respectively.

(i) Using the distributions of X and Y *only*, calculate

(a) $\text{P}(X + Y = 0)$

(b) $\text{P}(X + Y = 1)$

(c) $\text{P}(X + Y = 2)$.

The random variable T is defined by $T = X + Y$.

(ii) Write down the distribution of T.

(iii) Use your distribution from part (ii) to check your results in part (i).

M **4** A boy is watching vehicles travelling along a motorway. All the vehicles he sees are either cars or lorries; the numbers of each may be modelled by two independent Poisson distributions. The mean number of cars per minute is 8.3 and the mean number of lorries per minute is 4.7.

(i) For a given period of one minute, find the probability that he sees

(a) exactly seven cars

(b) at least three lorries.

(ii) Calculate the probability that he sees a total of exactly ten vehicles in a given one-minute period.

(iii) Find the probability that he observes fewer than eight vehicles in a given period of 30 seconds.

M **5** The number of cats rescued by an animal shelter each day may be modelled by a Poisson distribution with parameter 2.5, while the number of dogs rescued each day may be modelled by an independent Poisson distribution with parameter 3.2.

(i) Calculate the probability that on a randomly chosen day the shelter rescues

(a) exactly two cats

(b) exactly three dogs

(c) exactly five cats and dogs in total.

(ii) Given that one day exactly five cats and dogs were rescued, find the conditional probability that exactly two of these animals were cats.

M **6** The numbers of emissions per minute from two radioactive substances, A and B, are independent and have Poisson distributions with means 2.8 and 3.25 respectively.

Find the probabilities that in a period of one minute there will be

(i) at least three emissions from substance A

(ii) one emission from one of the two substances and two emissions from the other substance

(iii) a total of five emissions.

M **7** At a coffee shop both hot and cold drinks are sold. The number of hot drinks sold per minute may be assumed to be a Poisson variable with mean 0.7 and the number of cold drinks sold per minute may be assumed to be an independent Poisson variable with mean 0.4.

(i) Calculate the probability that in a given one-minute period exactly one hot drink and one cold drink are sold.

(ii) Calculate the probability that in a given three-minute period fewer than three drinks altogether are sold.

(iii) In a given one-minute period exactly three drinks are sold. Calculate the probability that these are all hot drinks.

8 During a weekday, cars pass a census point on a quiet side road independently and at random times. The mean rate for westward travelling cars is two in any five-minute period, and for eastward travelling cars is three in any five-minute period.

Find the probability that

(i) there will be no cars passing the census point in a given two-minute period

(ii) at least one car from each direction will pass the census point in a given two-minute period

(iii) there will be exactly ten cars passing the census point in a given ten-minute period.

5.4 The Poisson approximation to the binomial distribution

Rare disease blights town

Chemical plant blamed

A rare disease is attacking residents of Avonford. In the last year alone five people have been diagnosed as suffering from it. This is over three times the national average.

The disease (known as *Palfrey's condition*) causes nausea and fatigue. One sufferer, James Louth (32), of Harpers Lane, has been unable to work for the last six months. His wife Muriel (29) said 'I am worried sick, James has lost his job and I am frightened that the children (Mark, 4, and Samantha, 2) will catch it.'

Mrs Louth blames the chemical complex on the industrial estate for the disease. 'There were never any cases before *Avonford Chemicals* arrived.'

Local environmental campaigner Roy James supports Mrs Louth. 'I warned the local council when planning permission was sought that this would mean an increase in this sort of illness. Normally we would expect 1 case in every 40 000 of the population in a year.'

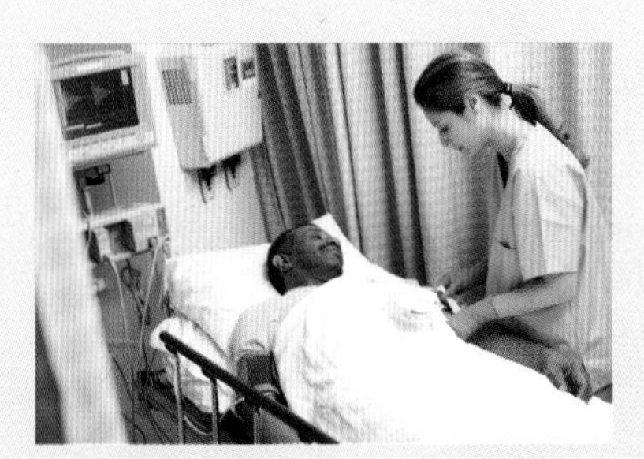

Muriel Louth believes that the local chemical plant could destroy her family's lives

Avonford Chemicals spokesperson Julia Millward said 'We categorically deny that our plant is responsible for the disease. Our record on safety is very good. None of our staff has had the disease. In any case five cases in a population of 60 000 can hardly be called significant.'

The expected number of cases is $60\,000 \times \frac{1}{40\,000}$ or 1.5, so 5 does seem rather high. Do you think that the chemical plant is to blame or do you think people are just looking for an excuse to attack it? How do you decide between the two points of view? Is 5 really that large a number of cases anyway?

The situation could be modelled by the binomial distribution. The probability of somebody getting the disease in any year is $\frac{1}{40\,000}$ and so that of not getting it is

$$1 - \frac{1}{40\,000} = \frac{39\,999}{40\,000}.$$

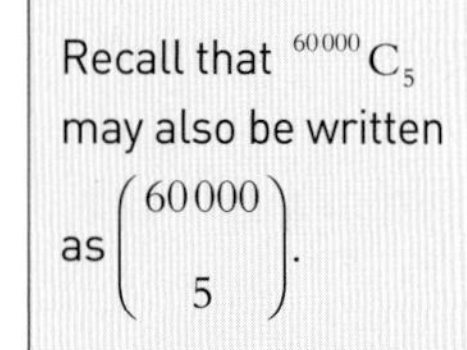

The probability of 5 cases among 60 000 people (and so 59 995 people not getting the disease) is given by

$$^{60\,000}C_5 \left(\frac{39\,999}{40\,000}\right)^{59\,995} \left(\frac{1}{40\,000}\right)^5 \approx 0.0141.$$

What you really want to know, however, is not the probability of exactly 5 cases but that of 5 or more cases. If that is very small, then perhaps something unusual did happen in Avonford last year.

You can find the probability of 5 or more cases by finding the probability of up to and including 4 cases, and subtracting it from 1.

The probability of up to and including 4 cases is given by:

$$\left(\frac{39\,999}{40\,000}\right)^{60\,000} \qquad \text{0 cases}$$

$$+ {}^{60\,000}C_1 \left(\frac{39\,999}{40\,000}\right)^{59\,999} \left(\frac{1}{40\,000}\right) \qquad \text{1 case}$$

$$+ {}^{60\,000}C_2 \left(\frac{39\,999}{40\,000}\right)^{59\,998} \left(\frac{1}{40\,000}\right)^2 \qquad \text{2 cases}$$

$$+ {}^{60\,000}C_3 \left(\frac{39\,999}{40\,000}\right)^{59\,997} \left(\frac{1}{40\,000}\right)^3 \qquad \text{3 cases}$$

$$+ {}^{60\,000}C_4 \left(\frac{39\,999}{40\,000}\right)^{59\,996} \left(\frac{1}{40\,000}\right)^4 \qquad \text{4 cases}$$

It is messy but you can evaluate it on your calculator. It comes out to be

$$0.223 + 0.335 + 0.251 + 0.126 + 0.047 = 0.981.$$

(The figures are written to three decimal places but more places were used in the calculation.)

So the probability of 5 or more cases in a year is $1 - 0.981 = 0.019$. It is unlikely but certainly could happen; see Figure 5.3 opposite.

Note

Two other points are worth making. First, the binomial model assumes the trials are independent. If this disease is at all infectious, that certainly would not be the case. Second, there is no evidence at all to link this disease with Avonford Chemicals. There are many other possible explanations.

▲ **Figure 5.3** Probability distribution $B\left(60\,000, \frac{1}{40\,000}\right)$

Approximating the binomial terms

Although it was possible to do the calculation using results derived from the binomial distribution, it was distinctly cumbersome. In this section, you will see how the calculations can be simplified, a process which turns out to be unexpectedly profitable. The work that follows depends upon the facts that the event is rare but there are many opportunities for it to occur: that is, p is small and n is large.

Start by looking at the first term, the probability of 0 cases of the disease. This is

$$\left(\frac{39\,999}{40\,000}\right)^{60\,000} = k, \text{ a constant.}$$

Now look at the next term, the probability of 1 case of the disease. This is

$$^{60\,000}C_1\left(\frac{39\,999}{40\,000}\right)^{59\,999} \times \left(\frac{1}{40\,000}\right)$$

$$= \frac{60\,000 \times \left(\frac{39\,999}{40\,000}\right)^{60\,000} \times \left(\frac{40\,000}{39\,999}\right)}{40\,000}$$

$$= k \times \frac{60\,000}{39\,999}$$

$$\approx k \times \frac{60\,000}{40\,000}$$

$$= k \times 1.5.$$

Now look at the next term, the probability of 2 cases of the disease. This is

$$^{60\,000}C_2 \times \left(\frac{39\,999}{40\,000}\right)^{59\,998} \times \left(\frac{1}{40\,000}\right)^2$$

$$= \frac{60\,000 \times 59\,999}{2 \times 1} \times \left(\frac{39\,999}{40\,000}\right)^{60\,000} \times \left(\frac{40\,000}{39\,999}\right)^2 \times \left(\frac{1}{40\,000}\right)^2$$

$$= \frac{k \times 60\,000 \times 59\,999}{2 \times 1 \times 39\,999 \times 39\,999}$$

$$\approx \frac{k \times 60\,000 \times 60\,000}{2 \times 40\,000 \times 40\,000}$$

$$= k \times \frac{(1.5)^2}{2}.$$

Proceeding in this way leads to the following probability distribution for the number of cases of the disease.

Number of cases	0	1	2	3	4	...
Probability	k	$k \times 1.5$	$k \times \frac{(1.5)^2}{2!}$	$k \times \frac{(1.5)^3}{3!}$	$k \times \frac{(1.5)^4}{4!}$	...

Since the sum of the probabilities = 1,

$$k + k \times 1.5 + k \times \frac{(1.5)^2}{2!} + k \times \frac{(1.5)^3}{3!} + k \times \frac{(1.5)^4}{4!} + \ldots = 1$$

$$k\left[1 + 1.5 + \frac{(1.5)^2}{2!} + \frac{(1.5)^3}{3!} + \frac{(1.5)^4}{4!} + \ldots\right] = 1$$

The terms in the square brackets form a well-known series in pure mathematics, the exponential series e^x.

$$e^x = 1 + x + \frac{x^2}{2!} + \frac{x^3}{3!} + \frac{x^4}{4!} + \ldots$$

Since $k \times e^{1.5} = 1$, $k = e^{-1.5}$.

This gives the probability distribution for the number of cases of the disease as

Number of cases	0	1	2	3	4	...
Probability	$e^{-1.5}$	$e^{-1.5} \times 1.5$	$e^{-1.5} \times \frac{(1.5)^2}{2!}$	$e^{-1.5} \times \frac{(1.5)^3}{3!}$	$e^{-1.5} \times \frac{(1.5)^4}{4!}$	...

and in general for r cases the probability is $e^{-1.5} \times \frac{(1.5)^r}{r!}$.

Accuracy

These expressions are clearly much simpler than those involving binomial coefficients. How accurate are they? The following table compares the results from the two methods, given to six decimal places.

Number of cases	Probability	
	Exact binomial method	**Approximate method**
0	0.223 126	0.223 130
1	0.334 697	0.334 695
2	0.251 025	0.251 021
3	0.125 512	0.125 511
4	0.047 066	0.047 067

You will see that the agreement is very good; there are no differences until the fifth or sixth decimal place.

The Poisson distribution may be used as an approximation to the binomial distribution, $B(n, p)$, when:

- n is large (typically $n > 50$)
- p is small (and so the event is rare)
- np is not too large (typically $np < 5$).

Example 5.5

It is known that nationally one person in a thousand is allergic to a particular chemical used in making a wood preservative. A firm that makes this wood preservative employs 500 people in one of its factories.

(i) What is the probability that more than two people at the factory are allergic to the chemical?
(ii) What assumption are you making?

Solution

(i) Let X be the number of people in a random sample of 500 who are allergic to the chemical.

$$X \sim B(500, 0.001) \qquad n = 500 \qquad p = 0.001$$

Since n is large and p is small, the Poisson approximation to the binomial is appropriate.

$$\begin{aligned}\lambda &= np \\ &= 500 \times 0.001 \\ &= 0.5\end{aligned}$$

Consequently $\quad P(X = r) = e^{-\lambda} \times \dfrac{\lambda^r}{r!}$

$$= e^{-0.5} \times \frac{0.5^r}{r!}$$

$$\begin{aligned}P(X > 2) &= 1 - P(X \leqslant 2) \\ &= 1 - [P(X = 0) + P(X = 1) + P(X = 2)] \\ &= 1 - \left[e^{-0.5} + e^{-0.5} \times 0.5 + e^{-0.5} \times \frac{0.5^2}{2}\right] \\ &= 1 - [0.6065 + 0.3033 + 0.0758] \\ &= 1 - 0.9856 \\ &= 0.0144\end{aligned}$$

→

(ii) The assumption made is that people with the allergy are just as likely to work in the factory as those without the allergy. In practice this seems rather unlikely: you would not stay in a job that made you unhealthy.

Exercise 5C

1 For each of the following binomial distributions, use the binomial formula to calculate $P(X = 3)$. Give your answers to 6 significant figures. *In each case* use an appropriate Poisson approximation to find $P(X = 3)$; again give your answers to 6 significant figures and calculate the percentage error in using this approximation. Describe what you notice.

(i) $X \sim B(25, 0.2)$

(ii) $X \sim B(250, 0.02)$

(iii) $X \sim B(2500, 0.002)$

M **2** An automatic machine produces washers, 3% of which are defective according to a severe set of specifications. A sample of 100 washers is drawn at random from the production of this machine. Using a suitable approximating distribution, calculate the probabilities of observing

(i) exactly 3 defectives

(ii) between 2 and 4 defectives inclusive.

M **3** The number of civil lawsuits filed in state and federal courts on a given day is 500. The probability that any such lawsuit is settled within one week is 0.01. Use the Poisson approximation to find the probability that, of the original 500 lawsuits on a given day, the number that are settled within a week is

(i) exactly seven

(ii) at least five

(iii) at most six.

4 One per cent of the items produced by a certain process are defective. Using the Poisson approximation, determine the probability that in a random sample of 400 articles

(i) exactly five are defective

(ii) at most five are defective.

M **5** Betty drives along a 50-kilometre stretch of road five days a week, 50 weeks a year. She takes no notice of the $70\,\text{km}\,\text{h}^{-1}$ speed limit and, when the traffic allows, travels between 95 and $105\,\text{km}\,\text{h}^{-1}$. From time to time she is caught by the police and fined but she estimates the probability of this happening on any day is $\frac{1}{300}$. If she gets caught three times within three years she will be disqualified from driving. Use Betty's estimates of probability to answer the following questions.

(i) What is the probability of her being caught exactly once in any year?

(ii) What is the probability of her being caught less than three times in three years?

(iii) What is the probability of her being caught exactly three times in three years?

Betty is in fact caught one day and decides to be somewhat cautious, reducing her normal speed to between 85 and 95 km h^{-1}. She believes this will reduce the probability of her being caught to $\frac{1}{500}$.

(iv) What is the probability that she is caught less than twice in the next three years?

M **6** Motorists in a particular part of Malaysia have a choice between a direct route and a one-way scenic detour. It is known that on average one in forty of the cars on the road will take the scenic detour. The road engineer wishes to do some repairs on the scenic detour. He chooses a time when he expects 100 cars an hour to pass along the road.

Find the probability that, in any one hour,

(i) no cars will turn on to the scenic detour

(ii) at most 4 cars will turn on to the scenic detour.

Between 10.30 am and 11.00 am it will be necessary to block the road completely.

(iii) What is the probability that no car will be delayed?

7 On average, 1 in 2500 adults has a certain medical condition.

(i) Use a suitable approximation to find the probability that, in a random sample of 4000 people, more than 3 have this condition.

(ii) In a random sample of n people, where n is large, the probability that none has the condition is less than 0.05. Find the smallest possible value of n.

Cambridge International AS & A Level Mathematics 9709 Paper 71 Q5 November 2015

5.5 Using the normal distribution as an approximation for the Poisson distribution

You may use the normal distribution as an approximation for the Poisson distribution, provided that its parameter (mean) λ is sufficiently large for the distribution to be reasonably symmetrical and not positively skewed.

As a working rule λ should be at least 15.

> The letter μ is also commonly used in place of λ for the Poisson parameter.

If $\lambda = 15$, mean = 15

and standard deviation = $\sqrt{15} = 3.87$ (to 3 s.f.).

A normal distribution is almost entirely contained within 3 standard deviations of its mean and in this case the value 0 is between 3 and 4 standard deviations away from the mean value of 15.

The parameters for the normal distribution are then

Mean:	$\mu = \lambda$
Variance:	$\sigma^2 = \lambda$

so that it can be denoted by $N(\lambda, \lambda)$.

(Remember that, for a Poisson distribution, mean = variance.)

For values of λ larger than 15 the Poisson probability graph becomes less positively skewed and more bell-shaped in appearance, thus making the normal approximation appropriate. Figure 5.4 shows the Poisson probability graph for the two cases $\lambda = 3$ and $\lambda = 25$. You will see that for $\lambda = 3$ the graph is positively skewed but for $\lambda = 25$ it is approximately bell-shaped.

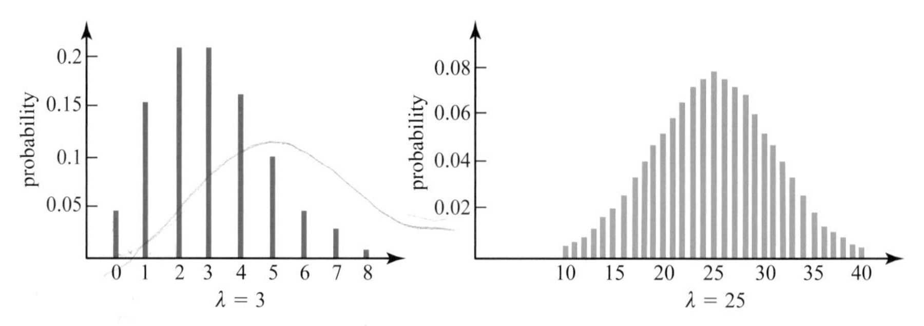

▲ **Figure 5.4**

Example 5.6

The annual number of deaths nationally from a rare disease, X, may be modelled by the Poisson distribution with mean 25. One year there are 31 deaths and it is suggested that the disease is on the increase.

What is the probability of 31 or more deaths in a year, assuming the mean has remained at 25?

Solution

The Poisson distribution with mean 25 may be approximated by the normal distribution with parameters

Mean: 25
Standard deviation: $\sqrt{25} = 5$

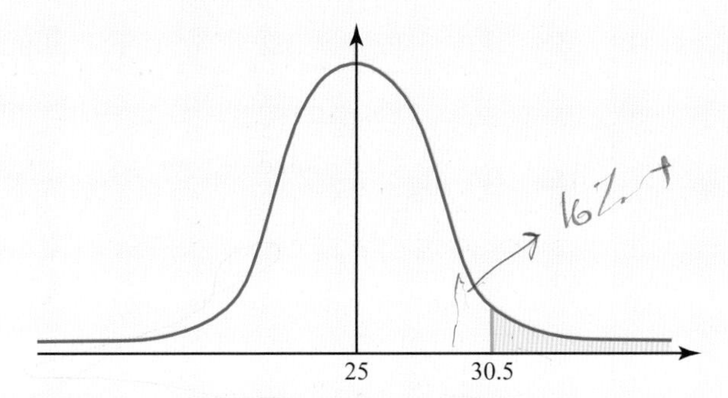

▲ **Figure 5.5**

The probability of there being 31 or more deaths in a year, $P(X \geqslant 31)$, is given by $1 - \Phi(z)$, where

$$z = \frac{30.5 - 25}{5} = 1.1$$

(Note the continuity correction, replacing 31 by 30.5.)

The required area is

$$1 - \Phi(1.1) = 1 - 0.8643$$
$$= 0.1357$$

This is not a particularly low probability; it is quite likely that there would be that many deaths in any one year.

5.6 Hypothesis test for the mean of a Poisson distribution

The next example shows you how to carry out a hypothesis test for the mean of a Poisson distribution.

Example 5.7

An old university has a high tower that is quite often struck by lightning. Records going back over hundreds of years show that on average the tower is struck on 3.2 days per year.

It is suggested that a likely effect of global warming would be an increase in the number of days on which the tower is struck. The following year the tower is struck by lightning on 7 days.

Carry out a suitable hypothesis test at the 5% significance level and state your conclusion.

What is the probability of a Type I error in this test?

Solution

The number of days per year that the tower is struck by lightning is modelled by X where

$$X \sim \text{Po}(3.2).$$

So the null and alternative hypotheses may be stated as follows.

$H_0: \mu = 3.2$ The population mean, μ, is unchanged.

$H_1: \mu > 3.2$ The population mean, μ, has increased.

One-tailed test

Significance level: 0.05

The test is one-tailed because it is for an *increase* in lightning strikes. A test for a change would be two-tailed.

The probability of 7 or more days with lightning strikes is (1 – the probability of 6 or fewer such days).

The calculation is shown in the table.

Number of strikes	Probability
0	$e^{-3.2} = 0.04076...$
1	$3.2 \times e^{-3.2} = 0.13043...$
2	$\frac{3.2^2}{2!} \times e^{-3.2} = 0.20870...$
3	$\frac{3.2^3}{3!} \times e^{-3.2} = 0.22261...$
4	$\frac{3.2^4}{4!} \times e^{-3.2} = 0.17809...$
5	$\frac{3.2^5}{5!} \times e^{-3.2} = 0.11397...$
6	$\frac{3.2^6}{6!} \times e^{-3.2} = 0.06078...$
Total	$\mathbf{P(X \leqslant 6) = 0.95538...}$

So the probability that $X \geqslant 7$ is

$$1 - 0.95538... = 0.04461... = 0.045 \quad (3 \text{ d.p.})$$

Since $0.045 < 0.05$ (the significance level), the null hypothesis is rejected in favour of the alternative hypothesis at the 5% significance level.

The evidence supports the hypothesis that there has been an increase in the incidence of lightning strikes.

A Type I error occurs when a true null hypothesis is rejected. In this test, the rejection region is $X > 6$ and the probability of such a result is 0.04461... if $\mu = 3.2$. So the probability of a Type I error is 0.0446 (to 3 s.f.).

Note

Notice that the result in Example 5.7 does not *prove* that there has been an increase in the incidence of lightning strikes; it does, however, suggest that this may well be the case. The test was about lightning strikes and so in itself says nothing about global warming. Whether global warming is connected to the incidence of lightning strikes is not what was being tested; it formed no part of either the null or the alternative hypothesis.

Exercise 5D

1 The number of night calls to a fire station in a small town can be modelled by a Poisson distribution with a mean of 4.2 per night.

Use the normal approximation to the Poisson distribution to estimate the probability that in any particular week (Sunday to Saturday inclusive) the number of night calls to the fire station will be

(i) at least 30 (ii) exactly 30

(iii) between 25 and 35 inclusive.

M 2 At a busy intersection of roads, accidents requiring the summoning of an ambulance occur with a frequency, on average, of 1.8 per week. These accidents occur randomly, so that it may be assumed that they follow a Poisson distribution.

Use a suitable approximating distribution to find the probability that in any particular year (of 52 weeks) the number of accidents at the intersection will be

(i) at most 100 (ii) exactly 100

(iii) between 95 and 105 inclusive.

M 3 Tina is a traffic warden. The number of parking tickets she issues per day, from Monday to Saturday inclusive, may be modelled by a Poisson distribution with mean 11.5. By using suitable approximating distributions, find the probability that

(i) on a particulaı Tuesday she issues at least 15 parking tickets

(ii) during any week (excluding Sunday) she issues at least 50 parking tickets

(iii) during four consecutive weeks she issues

(a) at least 50 parking tickets each week

(b) at least 200 parking tickets altogether.

Account for the difference in the two answers.

M 4 The number of emails I receive per day on my computer may be modelled by a Poisson distribution with mean 8.5.

(i) Use the most appropriate method to calculate the probability that I will receive

(a) at least 8 emails tomorrow

(b) at least 240 emails next June.

(ii) What assumption do you have to make to find the probability in part (i) (b)?

(iii) Compare your answers to parts (i) (a) and (b) and account for the variation.

5 At a petrol station cars arrive independently and at random times at constant average rates of 8 cars per hour travelling east and 5 cars per hour travelling west.

(i) Find the probability that, in a quarter-hour period,

(a) one or more cars travelling east and one or more cars travelling west will arrive,

(b) a total of 2 or more cars will arrive.

(ii) Find the approximate probability that, in a 12-hour period, a total of more than 175 cars will arrive.

Cambridge International AS & A Level Mathematics 9709 Paper 7 Q6 June 2005

6 Some ancient documents from the pharaoh's astronomer are discovered in one of the pyramids. They include records, covering many years, of shooting stars during a certain part of one particular night of the year. The data are well modelled by a Poisson distribution with mean 5.6. A modern astronomer has a theory that there are now fewer shooting stars and so, on the right day and time, repeats the observation and carries out a suitable hypothesis test, using a 10% significance level.

(i) State the null and alternative hypotheses.

(ii) Find the rejection region for the test.

(iii) Find the probability of a Type I error.

The astronomer observes three shooting stars.

(iv) Carry out the hypothesis test.

7 A dressmaker makes dresses for Easifit Fashions. Each dress requires $2.5\,\text{m}^2$ of material. Faults occur randomly in the material at an average rate of 4.8 per $20\,\text{m}^2$.

(i) Find the probability that a randomly chosen dress contains at least 2 faults.

Each dress has a belt attached to it to make an outfit. Independently of faults in the material, the probability that a belt is faulty is 0.03. Find the probability that, in an outfit,

(ii) neither the dress nor its belt is faulty,

(iii) the dress has at least one fault and its belt is faulty.

The dressmaker attaches 300 randomly chosen belts to 300 randomly chosen dresses. An outfit in which the dress has at least one fault and its belt is faulty is rejected.

(iv) Use a suitable approximation to find the probability that fewer than 3 outfits are rejected.

Cambridge International AS & A Level Mathematics 9709 Paper 7 Q6 June 2006

8 It is proposed to model the number of people per hour calling a car breakdown service between the times 09 00 and 21 00 by a Poisson distribution.

(i) Explain why a Poisson distribution may be appropriate for this situation.

People call the car breakdown service at an average rate of 20 per hour, and a Poisson distribution may be assumed to be a suitable model.

(ii) Find the probability that exactly 8 people call in any half hour.

(iii) By using a suitable approximation, find the probability that exactly 250 people call in the 12 hours between 09 00 and 21 00.

Cambridge International AS & A Level Mathematics 9709 Paper 7 Q5 June 2007

9 Major avalanches can be regarded as randomly occurring events. They occur at a uniform average rate of 8 per year.

(i) Find the probability that more than 3 major avalanches occur in a 3-month period.

(ii) Find the probability that any two separate 4-month periods have a total of 7 major avalanches.

(iii) Find the probability that a total of fewer than 137 major avalanches occur in a 20-year period.

Cambridge International AS & A Level Mathematics 9709 Paper 7 Q3 June 2009

10 When a guitar is played regularly, a string breaks on average once every 15 months. Broken strings occur at random times and independently of each other.

(i) Show that the mean number of broken strings in a 5-year period is 4.

A guitar is fitted with a new type of string which, it is claimed, breaks less frequently. The number of broken strings of the new type was noted after a period of 5 years.

(ii) The mean number of broken strings of the new type in a 5-year period is denoted by λ. Find the rejection region for a test at the 10% significance level when the null hypothesis $\lambda = 4$ is tested against the alternative hypothesis $\lambda < 4$.

(iii) Hence calculate the probability of making a Type I error.

The number of broken guitar strings of the new type, in a 5-year period, was in fact 1.

(iv) State, with a reason, whether there is evidence at the 10% significance level that guitar strings of the new type break less frequently.

Cambridge International AS & A Level Mathematics 9709 Paper 7 Q5 June 2008

11 Every month Susan enters a particular lottery. The lottery company states that the probability, p, of winning a prize is 0.0017 each month. Susan thinks that the probability of winning is higher than this, and carries out a test based on her 12 lottery results in a one-year period. She accepts the null hypothesis $p = 0.0017$ if she has no wins in the year and accepts the alternative hypothesis $p > 0.0017$ if she wins a prize in at least one of the 12 months.

(i) Find the probability of the test resulting in a Type I error.

(ii) If in fact the probability of winning a prize each month is 0.0024, find the probability of the test resulting in a Type II error.

(iii) Use a suitable approximation, with $p = 0.0024$, to find the probability that in a period of 10 years Susan wins a prize exactly twice.

Cambridge International AS & A Level Mathematics
9709 Paper 7 Q5 November 2008

12 Pieces of metal discovered by people using metal detectors are found randomly in fields in a certain area at an average rate of 0.8 pieces per hectare. People using metal detectors in this area have a theory that ploughing the fields increases the average number of pieces of metal found per hectare. After ploughing, they tested this theory and found that a randomly chosen field of area 3 hectares yielded 5 pieces of metal.

(i) Carry out the test at the 10% level of significance.

(ii) What would your conclusion have been if you had tested at the 5% level of significance?

Jack decides that he will reject the null hypothesis that the average number is 0.8 pieces per hectare if he finds 4 or more pieces of metal in another ploughed field of area 3 hectares.

(iii) If the true mean after ploughing is 1.4 pieces per hectare, calculate the probability that Jack makes a Type II error.

Cambridge International AS & A Level Mathematics
9709 Paper 7 Q6 November 2006

13 A hospital patient's white blood cell count has a Poisson distribution. Before undergoing treatment the patient had a mean white blood cell count of 5.2. After the treatment a random measurement of the patient's white blood cell count is made, and is used to test at the 10% significance level whether the mean white blood cell count has decreased.

(i) State what is meant by a Type I error in the context of the question, and find the probability that the test results in a Type I error.

(ii) Given that the measured value of the white blood cell count after the treatment is 2, carry out the test.

(iii) Find the probability of a Type II error if the mean white blood cell count after the treatment is actually 4.1.

Cambridge International AS & A Level Mathematics
9709 Paper 71 Q7 June 2010

Historical note

Siméon Poisson was born in Pithiviers in France in 1781. Under family pressure he began to study medicine but after some time gave it up for his real interest, mathematics. For the rest of his life Poisson lived and worked as a mathematician in Paris. His contribution to the subject spanned a broad range of topics in both pure and applied mathematics, including integration, electricity and magnetism, and planetary orbits, as well as statistics. He was the author of between 300 and 400 publications and originally derived the Poisson distribution as an approximation to the binomial distribution.

When he was a small boy, Poisson had his hands tied by his nanny who then hung him from a hook on the wall so that he could not get into trouble while she went out. In later life he devoted a lot of time to studying the motion of a pendulum and claimed that this interest derived from his childhood experience of swinging against the wall.

KEY POINTS

1 **The Poisson probability distribution**

If $X \sim \text{Po}(\lambda)$, the parameter $\lambda > 0$.

$$\text{P}(X = r) = \text{e}^{-\lambda} \times \frac{\lambda^r}{r!} \qquad r \geqslant 0, r \text{ is an integer}$$

$$\text{E}(X) = \lambda$$

$$\text{Var}(X) = \lambda$$

2 **Conditions under which the Poisson distribution may be used**

The Poisson distribution is generally thought of as the probability distribution for the number of occurrences of a rare event.

Situations in which the mean number of occurrences is known (or can easily be found) but in which it is not possible, or even meaningful, to give values to n or p may be modelled using the Poisson distribution provided that the occurrences are:

- random
- independent.

3 **The sum of two Poisson distributions**

If $X \sim \text{Po}(\lambda)$, $Y \sim \text{Po}(\mu)$ and X and Y are independent

$$X + Y \sim \text{Po}(\lambda + \mu)$$

→

4 **Approximating to the binomial distribution**

The Poisson distribution may be used as an approximation to the binomial distribution, B(n, p), when:

- n is large (typically $n > 50$)
- p is small (and so the event is rare)
- np is not too large (typically $np < 5$).

It would be unusual to use the Poisson distribution with parameter, λ, greater than about 20.

5 **Using the normal distribution as an approximation for the Poisson distribution**

The Poisson distribution Po(λ) may be approximated by N(λ, λ), provided λ is about 15 or more.

LEARNING OUTCOMES

Now that you have finished this chapter, you should be able to

- recognise situations under which the Poisson distribution is likely to be an appropriate model
- use formulae to calculate probabilities for a Poisson distribution
- know and be able to use the mean and variance of a Poisson distribution
- know that the sum of two or more independent Poisson distributions is also a Poisson distribution
- recognise situations in which both the Poisson distribution and the binomial distribution might be appropriate models
- use the normal distribution as an approximation for the Poisson distribution, applying continuity corrections where appropriate
- carry out a hypothesis test for the mean of a Poisson distribution, including using the normal distribution as an approximation to test for the mean of a Poisson distribution
- calculate the probabilities of making Type I and Type II errors in specific situations involving tests based on a Poisson distribution.

6 Linear combinations of random variables

> To approach zero defects, you must have statistical control of processes.
> *David Wilson*

Unfair dismissal?

Janice

I've just had one of those days. Everything went wrong. First the school bus arrived 5 minutes late to pick up my little boy. Then it was wet and slippery and there were so many people about that I just couldn't walk at my normal speed; usually I take 15 minutes but that day it took me 18 to get to work. And then when I got to work I had to wait three and a half minutes for the lift instead of the usual half a minute. So instead of arriving my normal 10 minutes early I was one minute late.

Mrs Dickens just wouldn't listen. She said she did not employ people to make excuses and told me to leave there and then.

Do you think I have a case for unfair dismissal?

Like Janice, we all have days when everything goes wrong at once. There were three random variables involved in her arrival time at work: the time she had to wait for the school bus, S, the time she took to walk to work, W, and the time she had to wait for the lift, L.

Her total time for getting to work, T, was the sum of all three: $T = S + W + L$.

Janice's case was essentially that the probability of T taking such a large value was very small. To estimate that probability you would need information about the distributions of the three random variables involved. You would also need to know how to handle the sum of two or more (in this case three) random variables.

6.1 The expectation (mean) of a function of X, E(g[X])

Before you can do this, you need to extend some of the work you did in *Probability & Statistics 1* on random variables. There you learned that, for a discrete random variable X with $P(X = x_i) = p_i$,

its expectation $= E(X) = \mu = \sum x_i \times P(X = x_i) = \sum x_i p_i$

and its variance $= \sigma^2 = E[(X - \mu)^2] = \sum (x_i - \mu)^2 \times P(X = x_i) = \sum (x_i - \mu)^2 p_i$

$$= E(X^2) - [E(X)]^2 = \sum x_i^2 \times P(X = x_i) - \mu^2 = \sum x_i^2 p_i - \mu^2.$$

This only finds the expected value and variance of a particular random variable.

Sometimes you will need to find the mean, i.e. the expectation, of a function of a random variable. That sounds rather forbidding and you may think the same of the definition given below at first sight. However, as you will see in the next two examples, the procedure is straightforward and common sense.

» If g[X] is a function of the discrete random variable X then E(g[X]) is given by

$$E(g[X]) = \sum_i g[x_i] \times P(X = x_i).$$

Example 6.1

What is the expectation of the square of the number that comes up when a fair die is rolled?

Solution

Let the random variable X be the number that comes up when the die is rolled.

$$g[X] = X^2$$

$$E(g[X]) = E(X^2) = \sum_i x_i^2 \times P(X = x_i)$$

$$= 1^2 \times \tfrac{1}{6} + 2^2 \times \tfrac{1}{6} + 3^2 \times \tfrac{1}{6} + 4^2 \times \tfrac{1}{6} + 5^2 \times \tfrac{1}{6} + 6^2 \times \tfrac{1}{6}$$

$$= 1 \times \tfrac{1}{6} + 4 \times \tfrac{1}{6} + 9 \times \tfrac{1}{6} + 16 \times \tfrac{1}{6} + 25 \times \tfrac{1}{6} + 36 \times \tfrac{1}{6}$$

$$= \tfrac{91}{6}$$

$$= 15.17$$

Note:

This calculation could also have been set out in table form as shown below.

x	$P(X = x_i)$	x_i^2	$x_i^2 \times P(X = x_i)$
1	$\frac{1}{6}$	1	$\frac{1}{6}$
2	$\frac{1}{6}$	4	$\frac{4}{6}$
3	$\frac{1}{6}$	9	$\frac{9}{6}$
4	$\frac{1}{6}$	16	$\frac{16}{6}$
5	$\frac{1}{6}$	25	$\frac{25}{6}$
6	$\frac{1}{6}$	36	$\frac{36}{6}$
Total	1		$\frac{91}{6}$

So $E(g[X]) = \frac{91}{6} = 15.17$

› $E(X^2)$is not the same as $[E(X)]^2$. In this case $15.57 \neq 3.5^2$, which is 12.25. In fact, the difference between $E(X^2)$and $[E(X)]^2$ is very important in statistics. Why is this?

Example 6.2

A random variable X has the following probability distribution.

Outcome	1	2	3
Probability	0.4	0.4	0.2

(i) Calculate $E(4X + 5)$.

(ii) Calculate $4E(X) + 5$.

(iii) Comment on the relationship between your answers to parts (i) and (ii).

Solution

(i) $E(g[X]) = \sum_i g[x_i] \times P(X = x_i)$ with $g[X] = 4X + 5$

x_i	1	2	3
$g[x_i]$	9	13	17
$P(X = x_i)$	0.4	0.4	0.2

$$\begin{aligned} E(4X + 5) &= E(g[X]) \\ &= 9 \times 0.4 + 13 \times 0.4 + 17 \times 0.2 \\ &= 12.2 \end{aligned}$$

(ii) $E(X) = 1 \times 0.4 + 2 \times 0.4 + 3 \times 0.2 = 1.8$
and so

$$\begin{aligned} 4E(X) + 5 &= 4 \times 1.8 + 5 \\ &= 12.2 \end{aligned}$$

(iii) Clearly $E(4X + 5) = 4E(X) + 5$, both having the value 12.2.

6.2 Expectation: algebraic results

In Example 6.2 you found that $E(4X + 5) = 4E(X) + 5$.

The working was numerical, showing that both expressions came out to be 12.2, but it could also have been shown algebraically. This would have been set out as follows.

	Proof	*Reasons (general rules)*
$E(4X + 5)$	$= E(4X) + E(5)$	$E(X \pm Y) = E(X) \pm E(Y)$
	$= 4E(X) + E(5)$	$E(aX) = aE(X)$
	$= 4E(X) + 5$	$E(c) = c$

Look at the general rules on the right-hand side of the box above. (X and Y are random variables, a and c are constants.) They are important but they are also common sense.

Notice the last one, which in this case means the expectation of 5 is 5. Of course it is; 5 cannot be anything else but 5. It is so obvious that sometimes people find it confusing! In general

$$E(aX + c) = aE(X) + c$$

These rules can be extended to take in the expectation of the sum of two functions of a random variable.

$$E(f[X] + g[X]) = E(f[X]) + E(g[X])$$

where f and g are both functions of X.

Example 6.3

The random variable X has the following probability distribution.

x	1	2	3	4
$P(X = x)$	0.6	0.2	0.1	0.1

Find

(i) $Var(X)$
(ii) $Var(7)$
(iii) $Var(3X)$
(iv) $Var(3X + 7)$.

What general rule do the answers to parts (ii) and (iv) illustrate?

Solution

(i)

x	1	2	3	4
x^2	1	4	9	16
$P(X = x)$	0.6	0.2	0.1	0.1

$$\begin{aligned} E(X) &= 1 \times 0.6 + 2 \times 0.2 + 3 \times 0.1 + 4 \times 0.1 \\ &= 1.7 \\ E(X^2) &= 1 \times 0.6 + 4 \times 0.2 + 9 \times 0.1 + 16 \times 0.1 \\ &= 3.9 \\ \text{Var}(X) &= E(X^2) - [E(X)]^2 \\ &= 3.9 - 1.7^2 \\ &= 1.01 \end{aligned}$$

(ii) $$\begin{aligned} \text{Var}(7) &= E(7^2) - [E(7)]^2 \\ &= E(49) - [7]^2 \\ &= 49 - 49 \\ &= 0 \end{aligned}$$

General result
$\text{Var}(c) = 0$ for a constant c.
This result is obvious; a constant is constant and so can have no spread.

(iii) $$\begin{aligned} \text{Var}(3X) &= E[(3X)^2] - \mu^2 \\ &= E(9X^2) - [E(3X)]^2 \\ &= 9E(X^2) - [3E(X)]^2 \\ &= 9 \times 3.9 - (3 \times 1.7)^2 \\ &= 35.1 - 26.01 \\ &= 9.09 \end{aligned}$$

General result
$\text{Var}(aX) = a^2\,\text{Var}(X)$.
Notice that it is a^2 and not a on the right-hand side, but that taking the square root of each side gives the standard deviation of $(aX) = a \times$ standard deviation (X) as you would expect from common sense.

(iv) $$\begin{aligned} &\text{Var}(3X + 7) \\ &= E[(3X + 7)^2] \\ &\quad - [E(3X + 7)]^2 \\ &= E(9X^2 + 42X + 49) \\ &\quad - [3E(X) + 7]^2 \\ &= E(9X^2) + E(42X) + E(49) \\ &\quad - [3 \times 1.7 + 7]^2 \\ &= 9E(X^2) + 42E(X) \\ &\quad + 49 - 12.1^2 \\ &= 9 \times 3.9 + 42 \times 1.7 \\ &\quad + 49 - 146.41 \\ &= 9.09 \end{aligned}$$

General result
$\text{Var}(aX + c) = a^2\,\text{Var}(X)$.
Notice that the constant c does not appear on the right-hand side.

Exercise 6A

CP

1 The probability distribution of a random variable X is as follows.

x	1	2	3	4	5
$\mathrm{P}(X = x)$	0.1	0.2	0.3	0.3	0.1

(i) Find

(a) $\mathrm{E}(X)$

(b) $\mathrm{Var}(X)$.

(ii) Verify that $\mathrm{Var}(2X) = 4\mathrm{Var}(X)$.

CP

2 The probability distribution of a random variable X is as follows.

x	0	1	2
$\mathrm{P}(X = x)$	0.5	0.3	0.2

(i) Find

(a) $\mathrm{E}(X)$

(b) $\mathrm{Var}(X)$.

(ii) Verify that $\mathrm{Var}(5X + 2) = 25\mathrm{Var}(X)$.

CP

3 Prove that $\mathrm{Var}(aX - b) = a^2\mathrm{Var}(X)$ where a and b are constants.

4 The random variable X is the number of heads obtained when four unbiased coins are tossed. Construct the probability distribution for X and find

(i) $\mathrm{E}(X)$

(ii) $\mathrm{Var}(X)$

(iii) $\mathrm{Var}(3X + 4)$.

5 A discrete random variable W has the following distribution.

w	1	2	3	4	5	6
$\mathrm{P}(W = w)$	0.1	0.2	0.1	0.2	0.1	0.3

Find the mean and variance of

(i) $W + 7$

(ii) $6W - 5$.

6 A coin is biased so that the probability of obtaining a tail is 0.75. The coin is tossed four times and the random variable X is the number of tails obtained. Find

(i) $\mathrm{E}(2X)$

(ii) $\mathrm{Var}(3X)$.

CP

7 The discrete random variable X has probability distribution given by

$$P(X = x) = \frac{(4x + 7)}{68} \qquad \text{for } x = 1, 2, 3, 4.$$

(i) Find

(a) $E(X)$

(b) $E(X^2)$

(c) $E(X^2 + 5X - 2)$.

(ii) Verify that $E(X^2 + 5X - 2) = E(X^2) + 5E(X) - 2$.

PS

8 A bag contains four balls, numbered 10, 20, 30 and 40. One ball is chosen at random and the number, N, on the ball is noted. Each ball has an equal chance of being taken from the bag.

(i) Find $E(N)$ and $Var(N)$.

Two balls are chosen at random one after the other, with the first ball being replaced after it has been drawn. Let $\overline{N}$ be the arithmetic mean of the numbers on the two balls.

(ii) Find $E(\overline{N})$ and $Var(\overline{N})$.

6.3 The sums and differences of independent random variables

Sometimes, as in the case of Janice in the website forum thread on page 113, you may need to add or subtract a number of independent random variables. This process is illustrated in the next example.

Example 6.4

The possible lengths (in cm) of the blades of cricket bats form a discrete uniform distribution:

38, 40, 42, 44, 46.

The possible lengths (in cm) of the handles of cricket bats also form a discrete uniform distribution:

22, 24, 26.

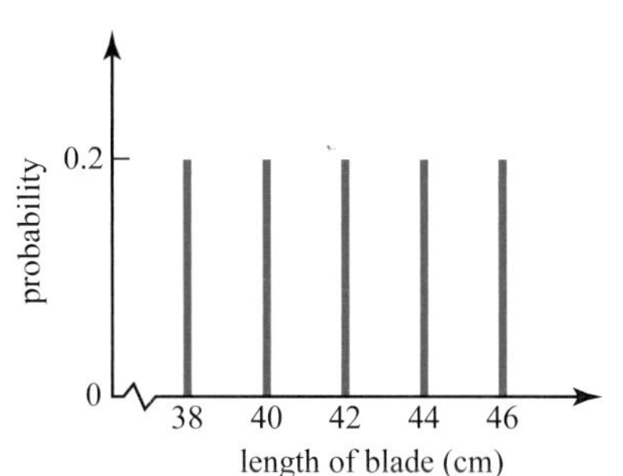

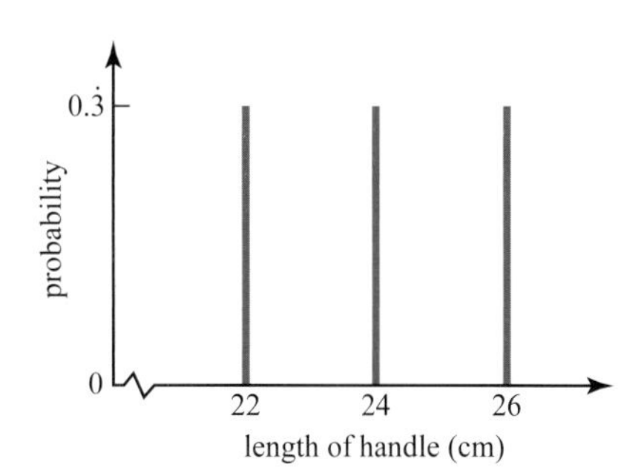

▲ **Figure 6.1**

The blades and handles can be joined together to make bats of various lengths, and it may be assumed that the lengths of the two sections are independent.

(i) How many different (total) bat lengths are possible?

(ii) Work out the mean and variance of random variable X_1, the length (in cm) of the blades.

(iii) Work out the mean and variance of random variable X_2, the length (in cm) of the handles.

(iv) Work out the mean and variance of random variable $X_1 + X_2$, the total length of the bats.

(v) Verify that

$$E(X_1 + X_2) = E(X_1) + E(X_2)$$

and

$$Var(X_1 + X_2) = Var(X_1) + Var(X_2).$$

Solution

(i) The number of different bat lengths is 7. This can be seen from the sample space diagram below.

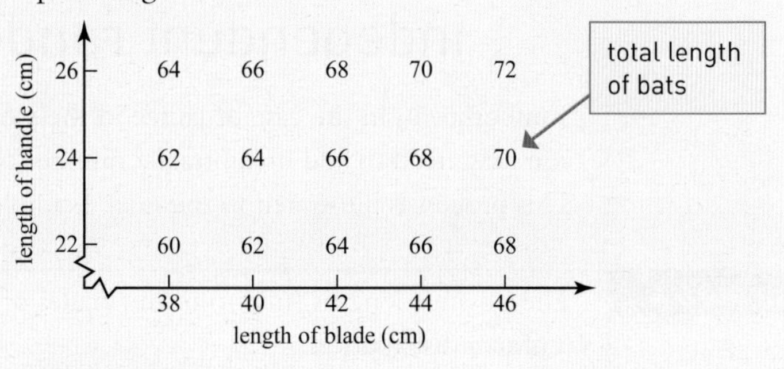

▲ **Figure 6.2**

(ii)

Length of blade (cm)	38	40	42	44	46
Probability	0.2	0.2	0.2	0.2	0.2

$$E(X_1) = \mu_1 = \sum xp$$
$$= (38 \times 0.2) + (40 \times 0.2) + (42 \times 0.2) + (44 \times 0.2) + (46 \times 0.2)$$
$$= 42 \text{ cm}$$
$$Var(X_1) = E(X_1^2) - \mu_1^2$$
$$E(X_1^2) = (38^2 \times 0.2) + (40^2 \times 0.2) + (42^2 \times 0.2) + (44^2 \times 0.2) + (46^2 \times 0.2)$$
$$= 1772$$
$$Var(X_1) = 1772 - 42^2 = 8$$

(iii)

Length of handle (cm)	22	24	26
Probability	$\frac{1}{3}$	$\frac{1}{3}$	$\frac{1}{3}$

$$\mathrm{E}(X_2) = \mu_2 = \left(22 \times \tfrac{1}{3}\right) + \left(24 \times \tfrac{1}{3}\right) + \left(26 \times \tfrac{1}{3}\right)$$
$$= 24\,\text{cm}$$
$$\mathrm{Var}(X_2) = \mathrm{E}(X_2^2) - \mu_2^2$$
$$\mathrm{E}(X_2^2) = \left(22^2 \times \tfrac{1}{3}\right) + \left(24^2 \times \tfrac{1}{3}\right) + \left(26^2 \times \tfrac{1}{3}\right)$$
$$= 578.667 \text{ to 3 d.p.}$$
$$\mathrm{Var}(X_2) = 578.667 - 24^2 = 2.667 \text{ to 3 d.p.}$$

(iv) The probability distribution of $X_1 + X_2$ can be obtained from Figure 6.2.

Total length of cricket bat (cm)	60	62	64	66	68	70	72
Probability	$\frac{1}{15}$	$\frac{2}{15}$	$\frac{3}{15}$	$\frac{3}{15}$	$\frac{3}{15}$	$\frac{2}{15}$	$\frac{1}{15}$

$$\mathrm{E}(X_1+X_2) = \left(60\times\tfrac{1}{15}\right)+\left(62\times\tfrac{2}{15}\right)+\left(64\times\tfrac{3}{15}\right)+\left(66\times\tfrac{3}{15}\right)+\left(68\times\tfrac{3}{15}\right)$$
$$+\left(70\times\tfrac{2}{15}\right)+\left(72\times\tfrac{1}{15}\right)$$
$$= 66 \text{ cm}$$
$$\mathrm{Var}(X_1+X_2) = \mathrm{E}[(X_1+X_2)^2] - 66^2$$
$$\mathrm{E}[(X_1+X_2)^2] = \left(60^2\times\tfrac{1}{15}\right)+\left(62^2\times\tfrac{2}{15}\right)+\left(64^2\times\tfrac{3}{15}\right)+\left(66^2\times\tfrac{3}{15}\right)+$$
$$\left(68^2\times\tfrac{3}{15}\right)+\left(70^2\times\tfrac{2}{15}\right)+\left(72^2\times\tfrac{1}{15}\right)$$
$$= \frac{65\,500}{15} = 4366.667 \text{ to 3 d.p.}$$
$$\mathrm{Var}(X_1+X_2) = 4366.667 - 66^2 = 10.667 \text{ to 3 d.p.}$$

(v) $\mathrm{E}(X_1 + X_2) = 66 = 42 + 24 = \mathrm{E}(X_1) + \mathrm{E}(X_2)$, as required.

$\mathrm{Var}(X_1 + X_2) = 10.667 = 8 + 2.667 = \mathrm{Var}(X_1) + \mathrm{Var}(X_2)$, as required.

Note

You should notice that the standard deviations of X_1 and X_2 do not add up to the standard deviation of $(X_1 + X_2)$.

$2.828 + 1.633 \neq \sqrt{10.667}$

i.e. $2.828 + 1.633 \neq 3.266$

General results

Example 6.4 has illustrated the following general results for the sums and differences of random variables.

For any two random variables X_1 and X_2:

» $\mathrm{E}(X_1 + X_2) = \mathrm{E}(X_1) + \mathrm{E}(X_2)$

Replacing X_2 by $-X_2$ in this result gives:

$$\mathrm{E}(X_1 + (-X_2)) = \mathrm{E}(X_1) + \mathrm{E}(-X_2)$$

» $\mathrm{E}(X_1 - X_2) = \mathrm{E}(X_1) - \mathrm{E}(X_2)$

If the variables X_1 and X_2 are independent then:

» $\text{Var}(X_1 + X_2) = \text{Var}(X_1) + \text{Var}(X_2)$

Replacing X_2 by $-X_2$ gives:

$$\text{Var}(X_1 + (-X_2)) = \text{Var}(X_1) + \text{Var}(-X_2)$$
$$\text{Var}(X_1 - X_2) = \text{Var}(X_1) + (-1)^2\,\text{Var}(X_2)$$

» $\text{Var}(X_1 - X_2) = \text{Var}(X_1) + \text{Var}(X_2)$

The sums and differences of normal variables

If the variables X_1 and X_2 have independent normal distributions, then the distributions of $(X_1 + X_2)$ and $(X_1 - X_2)$ are also normal. The means of these distributions are $\text{E}(X_1) + \text{E}(X_2)$ and $\text{E}(X_1) - \text{E}(X_2)$. Also $aX_1 + bX_2$ has a normal distribution.

You must, however, be careful when you come to their variances, since you may only use the result that

$$\text{Var}(X_1 \pm X_2) = \text{Var}(X_1) + \text{Var}(X_2)$$

to find the variances of these distributions if the variables X_1 and X_2 are independent.

This is the situation in the next two examples.

Example 6.5

Robert Fisher, a keen chess player, visits his local club most days. The total time taken to drive to the club and back is modelled by a normal variable with mean 25 minutes and standard deviation 3 minutes. The time spent at the chess club is also modelled by a normal variable with mean 120 minutes and standard deviation 10 minutes. Find the probability that on a certain evening Mr Fisher is away from home for more than $2\frac{1}{2}$ hours.

Solution

Let the random variable $X_1 \sim \text{N}(25, 3^2)$ represent the driving time, and the random variable $X_2 \sim \text{N}(120, 10^2)$ represent the time spent at the chess club.

Then the random variable T, where $T = X_1 + X_2 \sim \text{N}(145, (\sqrt{109})^2)$, represents his total time away.

So the probability that Mr Fisher is away for more than $2\frac{1}{2}$ hours (150 minutes) is given by

$$\text{P}(T > 150) = 1 - \Phi\left(\frac{150 - 145}{\sqrt{109}}\right)$$
$$= 1 - \Phi(0.479)$$
$$= 0.316.$$

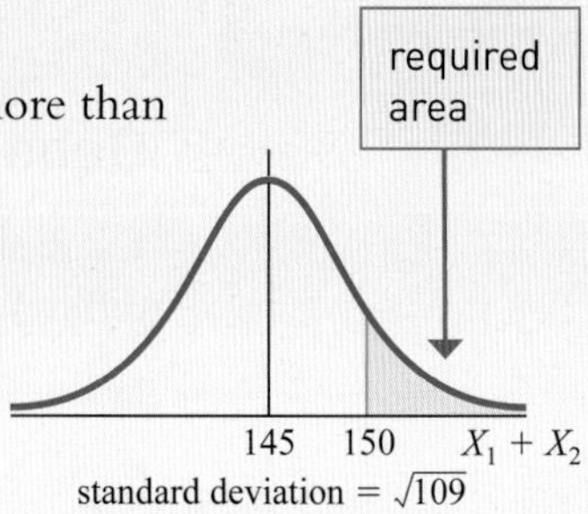

required area

145 150 $X_1 + X_2$

standard deviation $= \sqrt{109}$

▲ **Figure 6.3**

Example 6.6

In the manufacture of a bridge made entirely from wood, circular pegs have to fit into circular holes. The diameters of the pegs are normally distributed with mean 1.60 cm and standard deviation 0.01 cm, while the diameters of the holes are normally distributed with mean 1.65 cm and standard deviation 0.02 cm. What is the probability that a randomly chosen peg will not fit into a randomly chosen hole?

Solution

Let the random variable X be the diameter of a hole:

$$X \sim \mathrm{N}(1.65, 0.02^2) = \mathrm{N}(1.65, 0.0004).$$

Let the random variable Y be the diameter of a peg:

$$Y \sim \mathrm{N}(1.60, 0.01^2) = \mathrm{N}(1.6, 0.0001)$$

Let $F = X - Y$. F represents the gap remaining between the peg and the hole and so the sign of F determines whether or not a peg will fit in a hole.

$$\mathrm{E}(F) = \mathrm{E}(X) - \mathrm{E}(Y) = 1.65 - 1.60 = 0.05$$

$$\mathrm{Var}(F) = \mathrm{Var}(X) + \mathrm{Var}(Y) = 0.0004 + 0.0001 = 0.0005$$

$$F \sim \mathrm{N}(0.05, 0.0005)$$

If for any combination of peg and hole the value of F is negative, then the peg will not fit into the hole.

The probability that $F < 0$ is given by

$$\Phi\left(\frac{0 - 0.05}{\sqrt{0.0005}}\right) = 1 - \Phi(-2.236)$$

$$= 1 - 0.9873$$

$$= 0.0127.$$

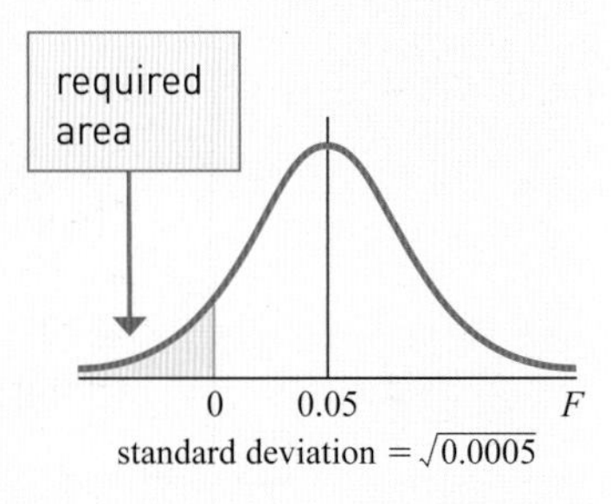

▲ **Figure 6.4**

Exercise 6B

CP

1 The menu at a café is shown below.

Main course		*Dessert*	
Fish and Chips	\$3	Ice Cream	\$1
Spaghetti	\$3.50	Apple Pie	\$1.50
Pizza	\$4	Sponge Pudding	\$2
Steak and Chips	\$5.50		

The owner of the café says that all the main-course dishes sell equally well, as do all the desserts, and that customers' choice of dessert is not influenced by the main course they have just eaten.

The variable M denotes the cost of the main course, in dollars, and the variable D the cost of the dessert. The variable T denotes the total cost of a two-course meal: $T = M + D$.

(i) Find the mean and variance of M.

(ii) Find the mean and variance of D.

(iii) List all the possible two-course meals, giving the price for each one.

(iv) Use your answer to part (iii) to find the mean and variance of T.

(v) Hence verify that for these figures

mean (T) = mean (M) + mean (D)

and variance (T) = variance (M) + variance (D).

2 X_1 and X_2 are independent random variables with distributions N(50, 16) and N(40, 9) respectively. Write down the distributions of

(i) $X_1 + X_2$

(ii) $X_1 - X_2$ → add for variance, subtract for mean

(iii) $X_2 - X_1$.

M **3** A play is enjoying a long run at a theatre. It is found that the play time may be modelled as a normal variable with mean 130 minutes and standard deviation 3 minutes, and that the length of the intermission in the middle of the performance may be modelled by a normal variable with mean 15 minutes and standard deviaton 5 minutes. Find the probability that the performance is completed in less than 140 minutes.

M **4** The time Melanie spends on her history assignments may be modelled as being normally distributed, with mean 40 minutes and standard deviation 10 minutes. The times taken on assignments may be assumed to be independent. Find

(i) the probability that a particular assignment will last longer than an hour

(ii) the time in which 95% of all assignments can be completed

(iii) the probability that two assignments will be completed in less than 75 minutes.

M **5** The weights of full cans of a particular brand of pet food may be taken to be normally distributed, with mean 260 g and standard deviation 10 g. The weights of the empty cans may be taken to be normally distributed, with mean 30 g and standard deviation 2 g. Find

(i) the mean and standard deviation of the weights of the contents of the cans

(ii) the probability that a full can weighs more than 270 g

(iii) the probability that two full cans together weigh more than 540 g.

6 The independent random variables X_1 and X_2 are distributed as follows:

$$X_1 \sim \mathrm{N}(30, 9); \quad X_2 \sim \mathrm{N}(40, 16).$$

Find the distributions of the following:

(i) $X_1 + X_2$

(ii) $X_1 - X_2$.

M **7** In a vending machine the capacity of cups is normally distributed, with mean 200 cm^3 and standard deviation 4 cm^3. The volume of coffee discharged per cup is normally distributed, with mean 190 cm^3 and standard deviation 5 cm^3. Find the percentage of drinks that overflow.

M **8** On a distant island the heights of adult men and women may both be taken to be normally distributed, with means 173 cm and 165 cm and standard deviations 10 cm and 8 cm respectively.

(i) Find the probability that a randomly chosen woman is taller than a randomly chosen man.

(ii) Do you think that this is equivalent to the probability that a married woman is taller than her husband?

M **9** The lifetimes of a certain brand of refrigerator are approximately normally distributed, with mean 2000 days and standard deviation 250 days. Mrs Chudasama and Mr Poole each buy one on the same date.

What is the probability that Mr Poole's refrigerator is still working one year after Mrs Chudasama's refrigerator has broken down?

PS **10** A random sample of size 2 is chosen from a normal distribution N(100, 25). Find the probability that

(i) the sum of the sample numbers exceeds 215

(ii) the first observation is at least 19 more than the second observation.

11 A mathematics module is assessed by an examination and by coursework. The examination makes up 75% of the total assessment and the coursework makes up 25%. Examination marks, X, are distributed with mean 53.2 and standard deviation 9.3. Coursework marks, Y, are distributed with mean 78.0 and standard deviation 5.1. Examination marks and coursework marks are independent. Find the mean and standard deviation of the combined mark $0.75X + 0.25Y$.

Cambridge International AS & A Level Mathematics 9709 Paper 7 Q2 June 2006

12 The cost of electricity for a month in a certain town under scheme A consists of a fixed charge of 600 cents together with a charge of 5.52 cents per unit of electricity used. Stella uses scheme A. The number of units she uses in a month is normally distributed with mean 500 and variance 50.41.

(i) Find the mean and variance of the total cost of Stella's electricity in a randomly chosen month.

Under scheme B there is no fixed charge and the cost in cents for a month is normally distributed with mean 6600 and variance 421. Derek uses scheme B.

(ii) Find the probability that, in a randomly chosen month, Derek spends more than twice as much as Stella spends.

Cambridge International AS & A Level Mathematics 9709 Paper 7 Q4 November 2007

6.4 More than two independent random variables

The results on pages 121–122 may be generalised to give the mean and variance of the sums and differences of n random variables, $X_1, X_2, \ldots, X_n$.

» $\mathrm{E}(X_1 \pm X_2 \pm \ldots \pm X_n) = \mathrm{E}(X_1) \pm \mathrm{E}(X_2) \pm \ldots \pm \mathrm{E}(X_n)$

and, provided $X_1, X_2, \ldots, X_n$ are independent,

» $\mathrm{Var}(X_1 \pm X_2 \pm \ldots \pm X_n) = \mathrm{Var}(X_1) + \mathrm{Var}(X_2) + \ldots + \mathrm{Var}(X_n)$.

If $X_1, X_2, \ldots, X_n$ is a set of normally distributed variables, then the distribution of $(X_1 \pm X_2 \pm \ldots \pm X_n)$ is also normal.

Example 6.7

The mass, X, of a suitcase at an airport is modelled as being normally distributed, with mean 15 kg and standard deviation 3 kg. Find the probability that a random sample of ten suitcases weighs more than 154 kg.

Solution

The mass X of one suitcase is given by

$$X \sim \mathrm{N}(15, 9).$$

Then the mass of each of the ten suitcases has the distribution of X; call them $X_1, X_2, \ldots, X_{10}$.

Let the random variable T be the total weight of ten suitcases.

$$T = X_1 + X_2 + \ldots + X_{10}.$$

$$\mathrm{E}(T) = \mathrm{E}(X_1) + \mathrm{E}(X_2) + \ldots + \mathrm{E}(X_{10})$$

$$= 15 + 15 + \ldots + 15$$

$$= 150$$

Similarly $\quad \text{Var}(T) = \text{Var}(X_1) + \text{Var}(X_2) + \ldots + \text{Var}(X_{10})$

$$= 9 + 9 + \ldots + 9$$
$$= 90$$

So $T \sim \text{N}(150, 90)$.

The probability that T exceeds 154 is given by

$$1 - \Phi\left(\frac{154 - 150}{\sqrt{90}}\right) = 1 - \Phi(0.422)$$
$$= 1 - 0.6635$$
$$= 0.3365.$$

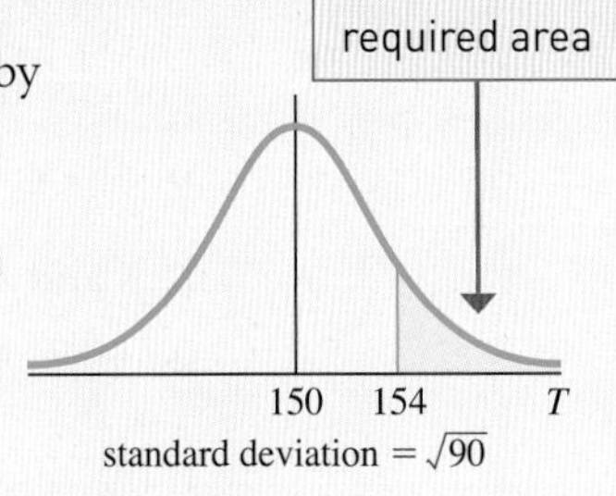

▲ **Figure 6.5**

Example 6.8

The running times of the four members of a 4 × 400 m relay team may all be taken to be normally distributed, as follows.

Member	Mean time (s)	Standard deviation (s)
Adil	52	1
Ben	53	1
Colin	55	1.5
Dexter	51	0.5

Assuming that no time is lost during changeovers, find the probability that the team finishes the race in less than 3 minutes 28 seconds.

Solution

Let the total time be T.

$$\text{E}(T) = 52 + 53 + 55 + 51 = 211$$
$$\text{Var}(T) = 1^2 + 1^2 + 1.5^2 + 0.5^2 = 4.5$$

So $T \sim \text{N}(211, 4.5)$.

The probability of a total time of less than 3 minutes 28 seconds (208 seconds) is given by

$$\Phi\left(\frac{208 - 211}{\sqrt{4.5}}\right) = \Phi(-1.414)$$
$$= 1 - 0.9213$$
$$= 0.0787.$$

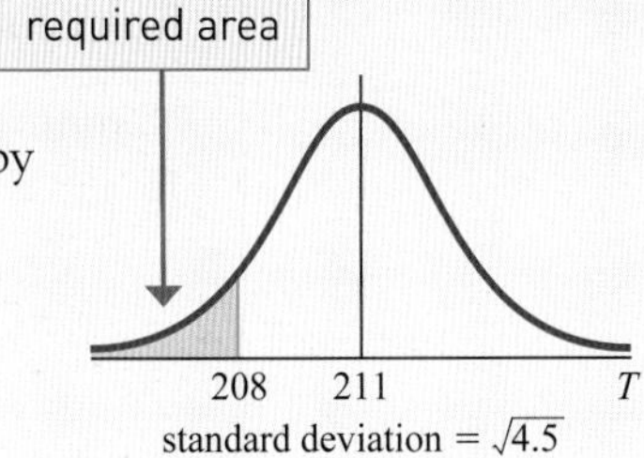

▲ **Figure 6.6**

Linear combinations of two or more independent random variables

The results given on pages 121–122 can also be generalised to include linear combinations of random variables.

For any random variables X and Y,

» $E(aX + bY) = aE(X) + bE(Y)$, where a and b are constants.

If X and Y are independent

» $\text{Var}(aX + bY) = a^2\text{Var}(X) + b^2\text{Var}(Y)$.

If the distributions of X and Y are normal, then the distribution of $(aX + bY)$ is also normal.

These results may be extended to any number of random variables.

Example 6.9

In a workshop, joiners cut out rectangular sheets of laminated board, of length L cm and width W cm, to be made into work surfaces. Both L and W may be taken to be normally distributed with standard deviation 1.5 cm. The mean of L is 150 cm, that of W is 60 cm, and the lengths L and W are independent. Both of the short sides and one of the long sides have to be covered by a protective strip (the other long side is to lie against a wall and so does not need protection).

What is the probability that a protecting strip 275 cm long will be too short for a randomly selected work surface?

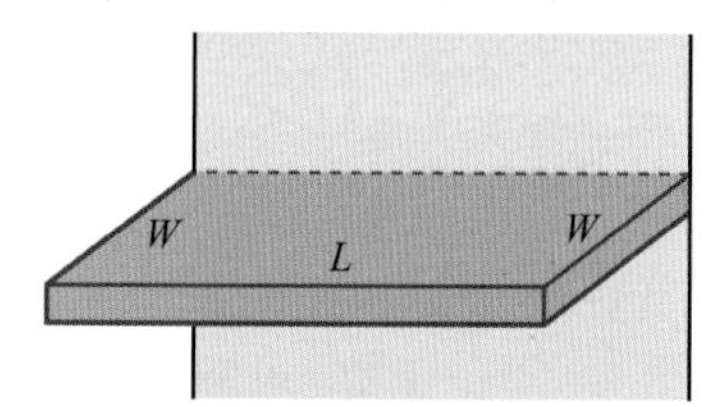

▲ **Figure 6.7**

Solution

Denoting the length and width by the independent random variables L and W and the total length of strip required by T

$$T = L + 2W$$

$$\begin{aligned}E(T) &= E(L) + 2E(W)\\ &= 150 + 2 \times 60\\ &= 270\end{aligned}$$

$$\begin{aligned}\text{Var}(T) &= \text{Var}(L) + 2^2\,\text{Var}(W)\\ &= 1.5^2 + 4 \times 1.5^2\\ &= 11.25.\end{aligned}$$

The probability of a strip 275 cm long being too short is given by

$$1 - \Phi\left(\frac{275 - 270}{\sqrt{11.25}}\right) = 1 - \Phi(1.491)$$
$$= 1 - 0.932$$
$$= 0.068.$$

Note

You have to distinguish carefully between the random variable $2W$, which means twice the size of one observation of the random variable W, and the random variable $W_1 + W_2$, which is the sum of two independent observations of the random variable W.

In the last example $\quad E(2W) = 2E(W) = 120$

and $\quad Var(2W) = 2^2 Var(W) = 4 \times 2.25 = 9.$

In contrast, $\quad E(W_1 + W_2) = E(W_1) + E(W_2) = 60 + 60 = 120$

and $\quad Var(W_1 + W_2) = Var(W_1) + Var(W_2) = 2.25 + 2.25 = 4.5.$

Example 6.10

A machine produces sheets of paper the thicknesses of which are normally distributed with mean 0.1 mm and standard deviation 0.006 mm.

(i) State the distribution of the total thickness of eight randomly selected sheets of paper.

(ii) Single sheets of paper are folded three times (to give eight thicknesses). State the distribution of the total thickness.

Solution

Denote the thickness of one sheet (in mm) by the random variable W, and the total thickness of eight sheets by T.

(i) *Eight separate sheets*

In this situation $T = W_1 + W_2 + W_3 + W_4 + W_5 + W_6 + W_7 + W_8$

where $W_1, W_2, \ldots, W_8$ are eight independent observations of the variable W. The distribution of W is normal with mean 0.1 and variance 0.006^2.

So the distribution of T is normal with

$$\text{mean} = 0.1 + 0.1 + \ldots + 0.1 = 8 \times 0.1 = 0.8$$

$$\text{variance} = 0.006^2 + 0.006^2 + \ldots + 0.006^2$$
$$= 8 \times 0.006^2$$
$$= 0.000\,288$$

$$\text{standard deviation} = \sqrt{0.000\,288} = 0.017.$$

The distribution is $N(0.8, 0.017^2)$.

→

(ii) *Eight thicknesses of the same sheet*

In this situation $T = W_1 + W_1 + W_1 + W_1 + W_1 + W_1 + W_1 + W_1 = 8W_1$ where W_1 is a single observation of the variable W.

So the distribution of T is normal with

$$\text{mean} = 8 \times \text{E}(W) = 0.8$$

$$\text{variance} = 8^2 \times \text{Var}(W) = 8^2 \times 0.006^2 = 0.002\,304$$

$$\text{standard deviation} = \sqrt{0.002\,304} = 0.048.$$

The distribution is $\text{N}(0.8, 0.048^2)$.

› Notice that in both cases the mean thickness is the same but for the folded paper the variance is greater. Why is this?

Exercise 6C

1 A garage offers motorists 'Road worthiness tests While U Wait' and claims that an average test takes only 20 minutes. Assuming that the time taken can be modelled as a normal variable with mean 20 minutes and standard deviation 2 minutes, find the distribution of the total time taken to conduct six tests in succession at this garage. State any assumptions you make.

2 A company manufactures floor tiles of mean length 20 cm with standard deviation 0.2 cm. Assuming the distribution of the lengths of the tiles is normal, find the probability that, when 12 randomly selected floor tiles are laid in a row, their total length exceeds 241 cm.

3 The masses of wedding cakes produced at a bakery are independent and may be modelled as being normally distributed with mean 4 kg and standard deviation 100 g. Find the probability that a set of eight wedding cakes has a total mass between 32.3 kg and 32.7 kg.

4 A random sample of 15 items is chosen from a normal population with mean 30 and variance 9. Find the probability that the sum of the variables in the sample is less than 440.

5 The distributions of four independent random variables X_1, X_2, X_3 and X_4 are N(7, 9), N(8, 16), N(9, 4) and N(10, 1) respectively.

Find the distributions of

(i) $X_1 + X_2 + X_3 + X_4$

(ii) $X_1 + X_2 - X_3 - X_4$

(iii) $X_1 - X_2 - X_3 + X_4$.

6 The distributions of X and Y are N(100, 25) and N(110, 36), and X and Y are independent. Find

(i) the probability that $8X + 2Y < 1000$

(ii) the probability that $8X - 2Y > 600$.

7 The distributions of the independent random variables A, B and C are $N(35, 9)$, $N(30, 8)$ and $N(35, 9)$. Write down the distributions of

(i) $A + B + C$

(ii) $5A + 4B$

(iii) $A + 2B + 3C$

(iv) $4A - B - 5C$.

8 The distributions of the independent random variables X and Y are $N(60, 4)$ and $N(90, 9)$. Find the probability that

(i) $X - Y < -35$

(ii) $3X + 5Y > 638$

(iii) $3X > 2Y$.

CP 9 If $X \sim N(60, 4)$ and $Y \sim N(90, 9)$ and X and Y are independent, find the probability that

(i) when one item is sampled from each population, the one from the Y population is more than 35 greater than the one from the X population

(ii) the sum of a sample consisting of three items from population X and five items from population Y exceeds 638

(iii) the sum of a sample of three items from population X exceeds that of two items from population Y.

(iv) Comment on your answers to questions 8 and 9.

M 10 The distribution of the weights of those rowing in a very large regatta may be taken to be normal with mean 80 kg and standard deviation 8 kg.

(i) What total weight would you expect 70% of randomly chosen crews of four rowers to exceed?

(ii) State what assumption you have made in answering this question and comment on whether you consider it reasonable.

M 11 The quantity of fuel used by a coach on a return trip of 200 km is modelled as a normal variable with mean 45 litres and standard deviation 1.5 litres.

(i) Find the probability that in nine return journeys the coach uses between 400 and 406 litres of fuel.

(ii) Find the volume of fuel that is 95% certain to be sufficient to cover the total fuel requirements for two return journeys.

M 12 Assume that the weights of men and women may be taken to be normally distributed, men with mean 75 kg and standard deviation 4 kg, and women with mean 65 kg and standard deviation 3 kg.

At a village fair, tug-of-war teams consisting of either five men or six women are chosen at random. The competition is then run on a knock-out basis, with teams drawn out of a hat. If in the first round a women's team is drawn against a men's team, what is the probability that the women's team is the heavier? State any assumptions you have made and explain how they can be justified.

M

13 The four runners in a relay team have individual times, in seconds, which are normally distributed, with means 12.1, 12.2, 12.3, 12.4, and standard deviations 0.2, 0.25, 0.3, 0.35 respectively. Find the probability that, in a randomly chosen race,

(i) the total time of the four runners is less than 48 seconds

(ii) runners 1 and 2 take longer than runners 3 and 4.

What assumption have you made and how realistic is the model?

14 The random variable X has the distribution $N(3.2, 1.2^2)$. The sum of 60 independent observations of X is denoted by S. Find $P(S > 200)$.

Cambridge International AS & A Level Mathematics 9709 Paper 7 Q2 June 2007

15 Weights of garden tables are normally distributed with mean 36 kg and standard deviation 1.6 kg. Weights of garden chairs are normally distributed with mean 7.3 kg and standard deviation 0.4 kg. Find the probability that the total weight of 2 randomly chosen tables is more than the total weight of 10 randomly chosen chairs.

Cambridge International AS & A Level Mathematics 9709 Paper 7 Q3 November 2008

16 A journey in a certain car consists of two stages with a stop for filling up with fuel after the first stage. The length of time, T minutes, taken for each stage has a normal distribution with mean 74 and standard deviation 7.3. The length of time, F minutes, it takes to fill up with fuel has a normal distribution with mean 5 and standard deviation 1.7. The length of time it takes to pay for the fuel is exactly 4 minutes. The variables T and F are independent and the times for the two stages are independent of each other.

(i) Find the probability that the total time for the journey is less than 154 minutes.

(ii) A second car has a fuel tank with exactly twice the capacity of the first car. Find the mean and variance of this car's fuel fill-up time.

(iii) This second car's time for each stage of the journey follows a normal distribution with mean 69 minutes and standard deviation 5.2 minutes. The length of time it takes to pay for the fuel for this car is also exactly 4 minutes. Find the probability that the total time for the journey taken by the first car is more than the total time taken by the second car.

Cambridge International AS & A Level Mathematics 9709 Paper 7 Q7 November 2005

KEY POINTS

1 For any discrete random variable X and constants a and c:

$\mathrm{E}(c) = c$

$\mathrm{E}(aX) = a\mathrm{E}(X)$

$\mathrm{E}(aX + c) = a\mathrm{E}(X) + c$

$\mathrm{Var}(c) = 0$

$\mathrm{Var}(aX) = a^2\mathrm{Var}(X)$

$\mathrm{Var}(aX + c) = a^2\mathrm{Var}(X)$.

2 For two random variables X and Y, whether independent or not, and constants a and b,

$\mathrm{E}(X \pm Y) = \mathrm{E}(X) \pm \mathrm{E}(Y)$

$\mathrm{E}(aX + bY) = a\mathrm{E}(X) + b\mathrm{E}(Y)$

and, if X and Y are independent,

$\mathrm{Var}(X \pm Y) = \mathrm{Var}(X) + \mathrm{Var}(Y)$

$\mathrm{Var}(aX \pm bY) = a^2\mathrm{Var}(X) + b^2\mathrm{Var}(Y)$.

3 For a set of n random variables, $X_1, X_2, \ldots, X_n$,

$\mathrm{E}(X_1 \pm X_2 \pm \ldots \pm X_n) = \mathrm{E}(X_1) \pm \mathrm{E}(X_2) \pm \ldots \pm \mathrm{E}(X_n)$

and, if the variables are independent,

$\mathrm{Var}(X_1 \pm X_2 \pm \ldots \pm X_n) = \mathrm{Var}(X_1) + \mathrm{Var}(X_2) + \ldots + \mathrm{Var}(X_n)$.

4 If random variables are independent and normally distributed, so are the sums, differences and other linear combinations of them.

LEARNING OUTCOMES

Now that you have finished this chapter, you should be able to

- use linear combinations of independent normal random variables in solving problems
- find and use the mean of any linear combination of random variables
- find and use the variance of any linear combination of independent random variables.

Answers

The questions, with the exception of those from past question papers, and all example answers that appear in this book were written by the authors. Cambridge Assessment International Education bears no responsibility for the example answers to questions taken from its past question papers which are contained in this publication.

Non-exact numerical answers should be given correct to three significant figures (or one decimal place for angles in degrees) unless a different level of accuracy is specified in the question. You should avoid rounding figures until reaching your final answer.

Chapter 1

? (Page 1)

See text that follows.

? (Page 4)

(i) The population is made up of the member of parliament's constituents. The sample is a part of that population of constituents. Without information relating to how the constituents' views were elicited, the views obtained seem to be biased towards those constituents who bother to write to their member of parliament.

(ii) The population is made up of households in Karachi. We are not told how the sample is chosen. Even if a random sample of households were chosen the views obtained are still likely to be biased as the interview timing excludes the possibility of obtaining views of most of those residents in employment.

(iii) The population is made up of black residents in Chicago. The sample is made up of black people (and possibly some white people as the areas are 'predominantly black') from a number of areas in Chicago.
The survey may be biased in two ways:

- the areas may not be representative of all residential areas and therefore of all black people living in Chicago and
- given that police officers are carrying out the survey they are unlikely to obtain negative views.

? (Page 5)

Each student is equally likely to be chosen but samples including two or more students from the same class are not permissible so not all samples are equally likely.

Yes

Activity 1.1 (Page 6)

There is no single answer since there are several ways you could use the given random numbers to generate the sample. This is one possible answer.

14592	12471	16718	2771	7107
16371	17775	2595	4598	16592

Exercise 1A (Page 8)

1 No, not every possible item has an equal chance of being selected. The sample is dependent on the first number produced. To produce a simple random sample the accountant should generate a random number for each item he is sampling.

2 (i) Simple random sample

(ii) $\frac{1}{25}$

3 Mr Jones could randomly select, say, four roads and then randomly select 15 houses from each road. Or Mr Jones could use a database of addresses and assign each address a number and then use a random number generator to select the addresses.

4 Teegan has not produced a random sample; she has only found out about her customer's preferences and not information to encourage new customers.

5 (i) Charlie is more likely to be chosen as P(one of each) = 0.5 and P(2 heads) = P(2 tails) = 0.25.

(ii) Choose Charlie if HT, Anthea if TH and Bill if HH. If the outcome is TT then rethrow both coins.

6 (i) Cluster sampling. Choose representative streets or areas and sample from these streets or areas.

(ii) Stratified sample. Identify routes of interest and randomly sample trains from each route.

(iii) Stratified sample. Choose representative areas in the town and randomly sample from each area as appropriate.

(iv) Stratified sample as in part (iii).

(v) Depends on method of data collection. If survey is, say, via a postal enquiry, then a random sample may be selected from a register of addresses.

(vi) Cluster sampling. Routes and times are chosen and a traffic sampling station is established to randomly stop vehicles to test tyres.

(vii) Cluster sampling. Areas are chosen and households are then randomly chosen.

(viii) Cluster sampling. A period (or periods) is chosen to sample and speeds are surveyed.

(ix) Cluster sampling. Meeting places for 18-year-olds are identified and samples of 18-year-olds are surveyed, probably via a method to maintain privacy. This might be a questionnaire to ascertain required information.

(x) Random sampling. The school student list is used as a sampling frame to establish a random sample within the school.

Chapter 2

? (Page 11)

See text that follows.

? (Page 13)

It is reasonable to regard the height of a wave as random. No two waves are exactly the same and in a storm some are much bigger than others.

Exercise 2A (Page 17)

1 (i) $k = \frac{2}{35}$

(ii)

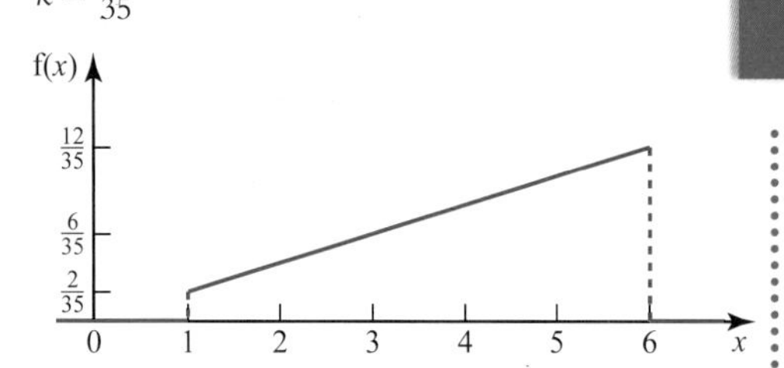

(iii) $\frac{11}{35}$

(iv) $\frac{1}{7}$

2 (i) $k = \frac{1}{12}$

(ii)

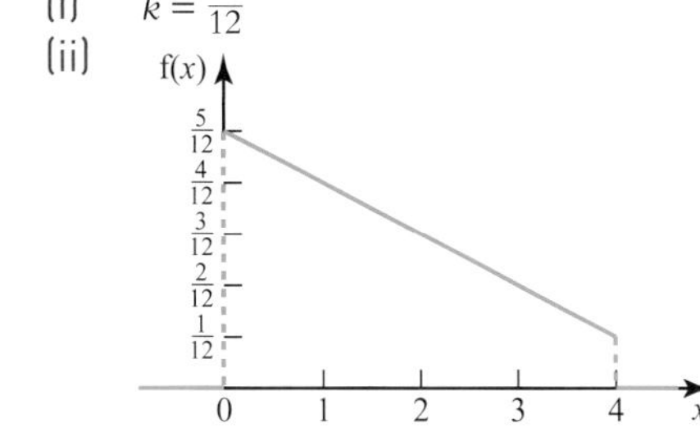

(iii) 0.207

3 (i) $a = \frac{4}{81}$

(ii)

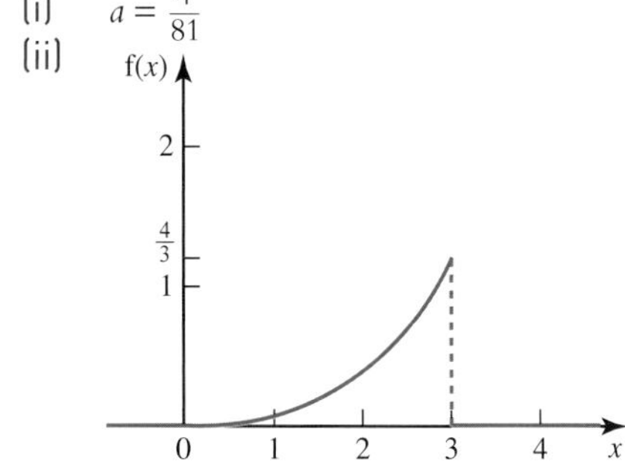

(iii) $\frac{16}{81}$

4 (i) $c = \frac{1}{8}$

(ii)

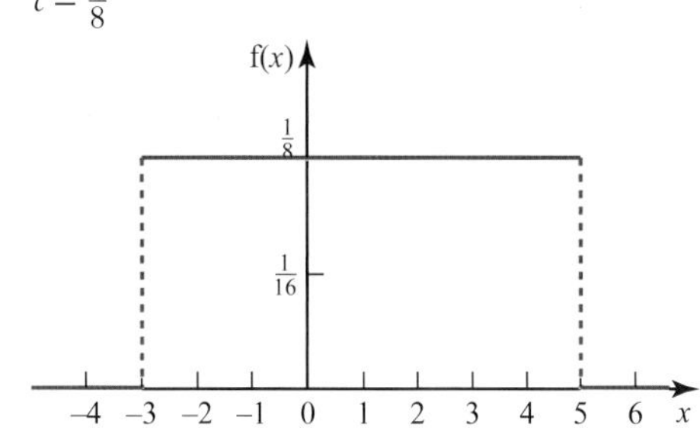

(iii) $\frac{1}{4}$

(iv) $\frac{3}{8}$

5 (i) $k = 0.048$

(ii)

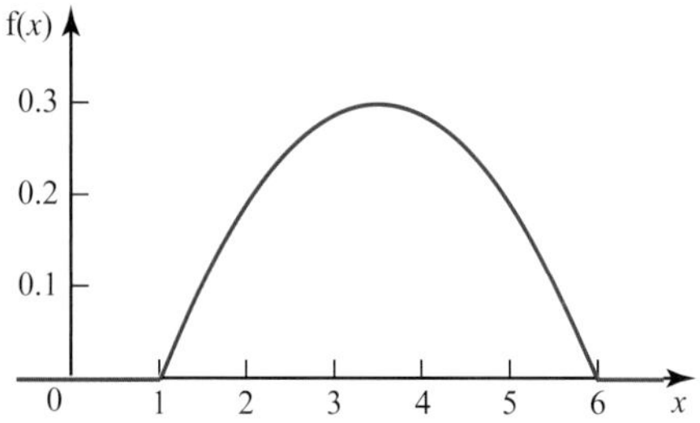

(iii) 0.248

6 (i) $k = \frac{2}{9}$

(ii) 0.0672

7 (i) $k = \frac{1}{100}$

(ii)

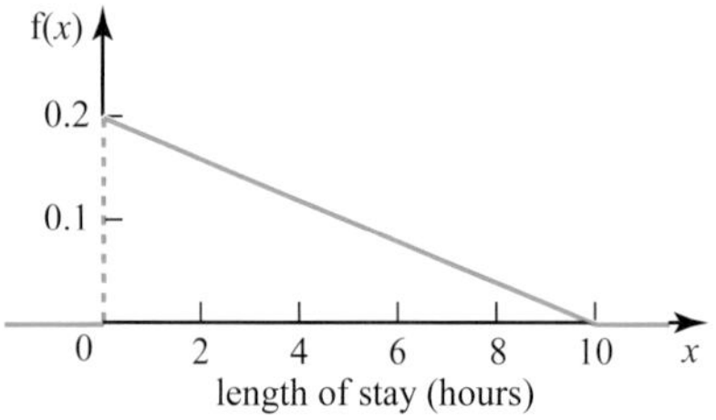

(iii) 19, 17, 28, 36

(iv) Yes

(v) Further information needed about the group 4–10 hours. It is possible that many of these stay all day and so are part of a different distribution.

8 (i)

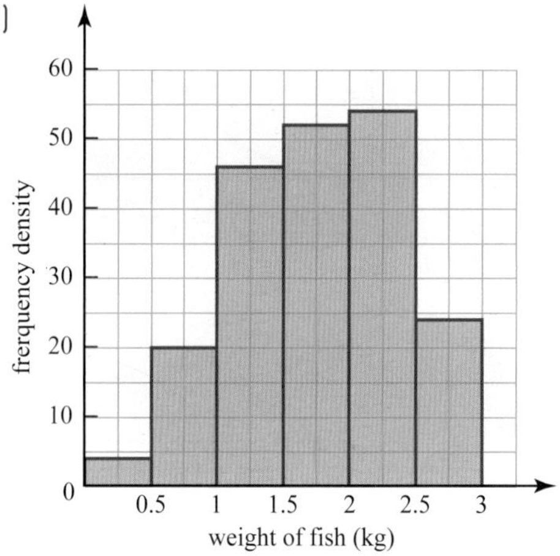

Negative skew

(ii)

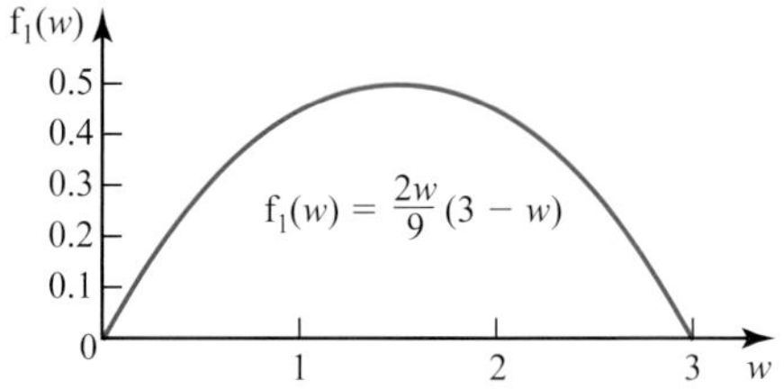

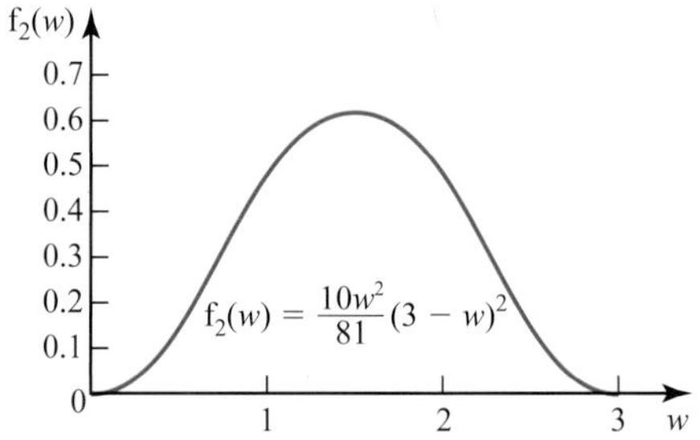

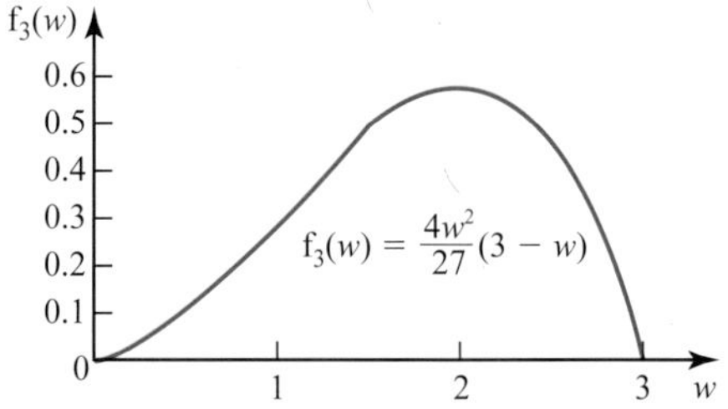

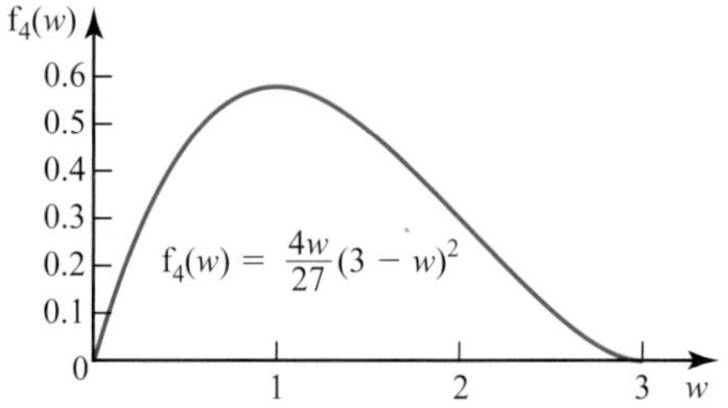

f_3 matches the data most closely in general shape.

(iii) 1.62, 9.49, 20.1, 28.0, 27.5, 13.2

(iv) Model seems good.

9 (ii) 0.916, 0.264

10 (i) 0, 0.1, 0.21, 0.12, 0.05, 0.02, 0

(ii) (a) 0.1 (b) 0.33 (c) 0.33
(d) 0.16 (e) 0.07 (f) 0.02

(iii) $k = \frac{1}{1728}$

(iv) (a) 0.132 (b) 0.275 (c) 0.280
(d) 0.201 (e) 0.0949 (f) 0.0162

(v) Model quite good. Both positively skewed.

11 (i) $k = 2$

(ii)

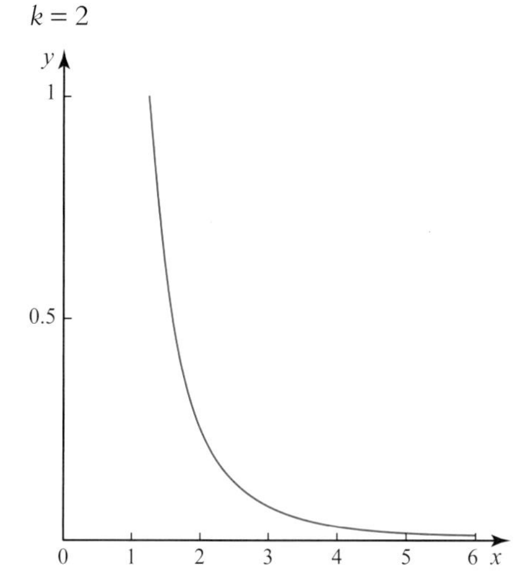

(iii) $m = 2$

12 (i) $k = \frac{1}{4}$

(ii) 10 years

(Page 24)

Distributions (ii) and (iv).

(Page 27)

68% (see *Probability & Statistics 1* Chapter 7). The normal distribution has a greater proportion of values near the mean, as can be seen from its shape.

Exercise 2B (Page 29)

1 (i) $2\frac{2}{3} = 2.67$

(ii) $\frac{8}{9} = 0.889$

(iii) $2\sqrt{2} = 2.83$

2 (i) 2 (ii) 2 (iii) 1.76

3 (i) 0.6

(ii) 0.04

4 (i)

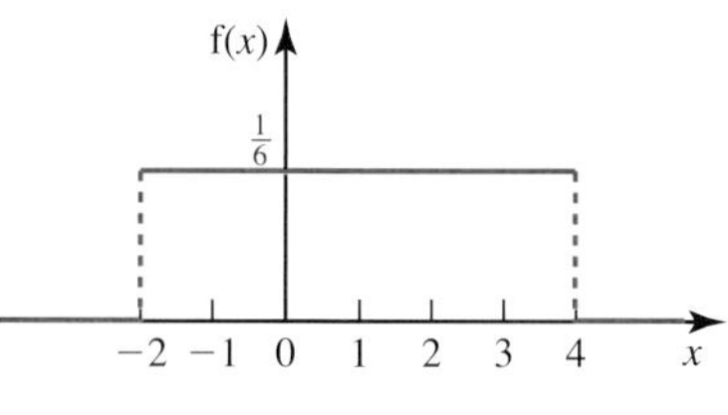

(ii) $\frac{2}{3}$ (iii) 1 (iv) $\frac{1}{3}$

5 (i) 1.5

(ii) 0.45

(iii) 1.5

(iv) 1.5

(v)

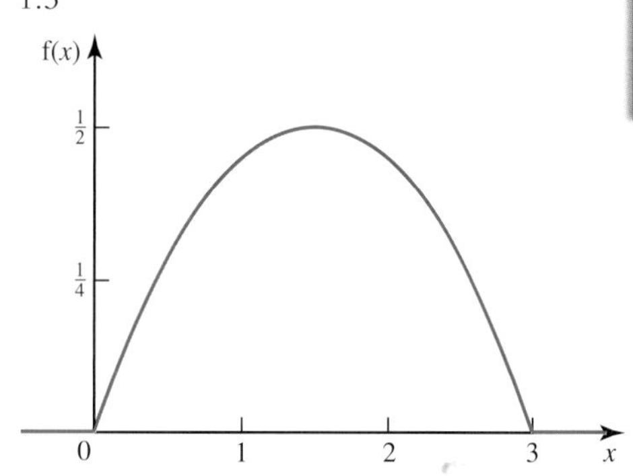

The graph is symmetrical and peaks when $x = 1.5$ thus $E(X) =$ mode of $X =$ median value of $X = 1.5$.

6 (ii) 1.08, 0.326

(iii) $\frac{9}{16} = 0.5625$

7 (i) $f(x) = \frac{1}{3}$ for $4 \leqslant x \leqslant 7$; $f(x) = 0$ otherwise

(ii) 5.5

(iii) $\frac{3}{4}$

(iv) 0.233

8 (i) $f(x) = \frac{1}{10}$ for $10 \leqslant x \leqslant 20$; $f(x) = 0$ otherwise

(ii) 15, 8.33

(iii) (a) 57.7%

(b) 100%

9 (i) $a = \frac{1}{\ln 2} = 1.44$

(ii)

(iii) 1.44, 0.0826

(iv) 41.5%

(v) $\sqrt{2} = 1.41$

10 (ii) 104 or 105 days

(iii) 5.14 hours

11 (ii) $\text{Var}(X) = 0.0267$

(iii) 0.931

(iv) 0.223

12 (ii) 0.139

(iii) 1.24 minutes

13 (ii) 4.125 minutes
(iii) 0.148
(iv) Less than 5 minutes, since $0.148 < 0.25$

14 (ii) 2.66 hours
(iii) 2.73 hours
(iv) 0.0241

15 (ii)

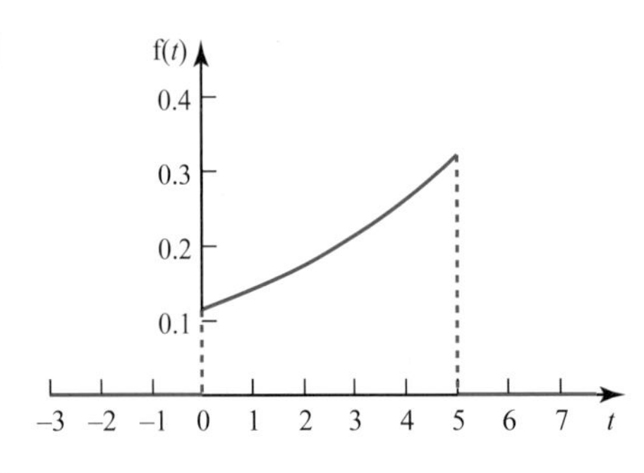

(iii) 1.48 seconds

Chapter 3

❓ (Page 34)

Medicines are only licensed after undergoing extensive trials supported by rigorous statistical tests.

❓ (Page 36)

Assuming both types of parents have the same fertility, boys born would outnumber girls in the ratio 3:1. In a generation's time there would be a marked shortage of women of child-bearing age.

❓ (Page 40)

Yes. The test was set up before the data were known, the data are random and independent and it is indeed testing the claim.

Exercise 3A (Page 41)

1 0.0573 Accept H_0

2 $0.0547 > 5\%$ Accept H_0

3 H_0: probability that toast lands butter-side down = 0.5
H_1: probability that toast lands butter-side down > 0.5
0.0154 Accept H_0

4 0.0474 Reject H_0
There is evidence that the complaints are justified at the 5% significance level, though Mr McTaggart might object that the candidates were not randomly chosen.

5 0.104 Accept H_0
Insufficient evidence at the 5% significance level that the machine needs servicing.

6 (i) 0.590
(ii) 0.0439
(iii) 0.0000712
(iv) 0.0292
(v) H_0: P(long question right) = 0.5
H_1: P(long question right) > 0.5
(vi) No

7 H_0: P(car is red) = 0.3
H_1: P(car is red) < 0.3
Accept H_0 (Isaac's claim) since $0.0600 > 5\%$

8 0.0834
Reject H_0
There is sufficient evidence at the 10% significance level that the player has improved.

❓ (Page 45)

Rejection region at 10% significance level is $X \leqslant 4$.

Exercise 3B (Page 46)

1 (i) $H_0: p = 0.9, H_1: p < 0.9$
(ii) (a) Number of parcels delivered the next day $\leqslant 12$.
(b) Number of parcels delivered the next day $\geqslant 13$.
(iii) 13 lies outside of the critical region, so insufficient evidence to say the service has deteriorated at the 5% significance level.

2 (i) $H_0: p = 0.5, H_1: p > 0.5$
(ii) (a) Number of even numbers $\geqslant 12$.
(b) Number of even numbers $\leqslant 11$.
(iii) Yes

3 (i) (a) 0.0991
(b) 0.139
(ii) $H_0: p = 0.8, H_1: p > 0.8$
(iii) Number of A to C passes $\geqslant 17$.
(iv) (a) 0.0991
(b) 0.861

Exercise 3C (Page 49)

1 $P(X \leqslant 3) = 0.0730 > 5\%$ Accept H_0

2 $P(X \geqslant 13) = 0.0106 < 2\frac{1}{2}\%$ Reject H_0

3 $P(X \geqslant 9) = 0.0730 > 2\frac{1}{2}\%$ Accept H_0

4 0 correct or > 6 correct

5 Critical region is $\leqslant 3$ or $\geqslant 13$ letter Zs

6 (i) 5
(ii) 0.196
(iii) Complaint justified at the 10% significance level.

Exercise 3D (Page 53)

1 (i) 4.80%
(ii) 0.0480
(iii) 0.601

2 (i) If H_0 is wrongly rejected because there were only 0 or 1 red chocolate beans in the sample although 20% of the population were actually red.
(ii) 0.167

3 (i) H_0: P(pass on 1st attempt) = 0.36
H_1: P(pass on 1st attempt) > 0.36
Reject H_0 (accept the driving instructor's claim) since 0.00429 < 5%
(ii) Type I error; 0.0293

4 (i) H_0: P(six) $= \frac{1}{6}$
H_1: P(six) $> \frac{1}{6}$
Accept H_0 since 0.225 > 10%
There is no evidence that the die is biased.
(ii) P(4 or more sixes) = 0.0697
(iii) Concluding that the die is fair when it is biased.

5 (i) 0 or 1 packet contain a gift
(ii) 0.0243
(iii) 2 is outside the rejection region. No evidence to reject claim.

6 (i) 0.256
(ii) 0.117
(iii) Type I; they will reject Luigi's theory, but it might be true.

Chapter 4

? (Page 57)

See text that follows.

Exercise 4A (Page 64)

1 (i) $z = 1.53$, not significant
(ii) $z = -2.37$, significant
(iii) $z = 1.57$, not significant
(iv) $z = 2.25$, significant
(v) $z = -2.17$, significant

2 (i) 0.309
(ii) 0.016
(iii) 0.00621
(iv) $H_0: \mu = 4.00$ g, $H_1: \mu > 4.00$ g
$z = 3$, significant

3 (i) $H_0: \mu = 72.7$ g,
$H_1: \mu \neq 72.7$ g; two-tailed test.
(ii) $z = 1.84$, not significant
(iii) No, significant

4 (i) $H_0: \mu = 23.9°$, $H_1: \mu > 23.9°$
(ii) $z = 1.29$, significant
(iii) 20.6; the standard deviation 4.54 is much greater than 2.3 so the ecologist should be asking whether the temperature has become more variable.

5 (i) You must assume that the visibilities are normally distributed.
(ii) $H_0: \mu = 14$ nautical miles,
$H_1: \mu < 14$ nautical miles
(iii) $z = -2.28$, significant
(iv) Choosing 36 consecutive days to collect data is not a good idea because weather patterns will ensure that the data are not independent. A better sampling procedure would be to choose every tenth day. In this way the effects of weather patterns over the year would be lessened.

6 (i) 999, 49.8
(ii) $H_0: \mu = 1000$, $H_1: \mu < 1000$
(iii) $z = -1.59$, not significant

7 $H_0: \mu = 0$, $H_1: \mu \neq 0$; $z = 0.983$, not significant

8 (i) 16.2, 27.4
(ii) $H_0: \mu = 15$, $H_1: \mu > 15$
(iii) $z = 1.99$, not significant

9 (i) 1.98, 0.0173
(ii) $H_0: \mu = 2$, $H_1: \mu < 2$
(iii) $z = -1.68$, not significant

10 (i) 105, 3.02
(ii) $H_0: \mu = 105, H_1: \mu \neq 105$
(iii) $z = -0.889$, not significant

11 (i) Commuters are not representative of the whole population.
(ii) Adults who travel to work on that train
(iii) Mean = 6.17 hours, variance = 0.657 hours

12 (i) $H_0: \mu = 21.2, H_1: \mu \neq 21.2$;
rejection region is $\bar{x} < 19.7$ or $\bar{x} > 22.7$;
$\bar{x} = 19.4$, significant ($19.4 < 19.7$)
There is significant evidence to suggest that the sentence length is not the same (or the book is not by the same author).
(ii) A Type I error would occur if you say that the sentence length is not the same (or the book is not by the same author) when it is.
Probability = 5%

13 (i) Mean = 6.525, variance = 2.87
(ii) $H_0: \mu = 7.2, H_1: \mu < 7.2$;
rejection region is $\bar{x} < 6.76$;
$\bar{x} = 6.525$, significant ($6.525 < 6.76$)
The evidence supports the hypothesis that there has been a reduction in the number of cars caught speeding.
(iii) A Type I error would occur if you say that there had been a reduction in the number of cars caught speeding when such a reduction had not occurred.

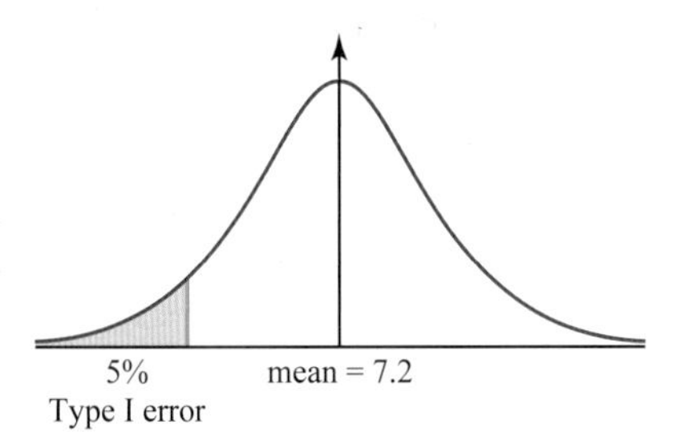

14 $H_0: \mu = 22.0, H_1: \mu \neq 22.0$;
rejection region is $\bar{x} < 21.698$;
$\bar{x} = 21.7$, not significant (just, $21.7 > 21.698$)
There is not enough evidence to say that the mean length has changed.

15 $H_0: \mu = 3.2, H_1: \mu < 3.2$, where μ is the growth rate of the new grass in cm per week
(i) $\bar{x} < 2.47$
(ii) $m < 2.47$

16 (i) $H_0: p = 0.2; H_1: p > 0.2$
(ii) Using the normal approximation to the binomial distribution,
$p(x > 12.5) = 0.188 > 0.1$, so the claim is not justified.

? (Page 70)

It tells you that μ is about 101.2 but it does not tell you what 'about' means, how close to 101.2 it is reasonable to expect μ to be.

? (Page 75)

You would expect about 90 out of the 100 to enclose 3.5.

Exercise 4B (Page 77)

1 (i) 5.205 g
(ii) 5.12 g, 5.29 g

2 (i) 47.7
(ii) 34.7 to 60.7
(iii) 27.3 to 68.1

3 (i) (a) 0.169
(b) 0.0207
(ii) 163.8–166.6 cm
(iii) 385

4 (i) 6.83, 3.04
(ii) 6.58, 7.08

5 (i) 5.71 to 7.49
(ii) It is more likely that the short manuscript was written in the early form of the language.

6 (i) 1.63 to 3.91
(ii) The coach's suspicions seem to be confirmed as 4 mm is not in the confidence interval.

7 0.484 to 1.02
Assume that the sample standard deviation is an acceptable approximation for σ.
The aim has not been achieved as the interval contains values below 0.5.

8 0.223 to 1.40
Assumptions: the Central Limit Theorem applies and s^2 is a good approximation for σ^2.
The confidence interval suggests that reaction times are slower after a meal.

9 (i) $\mu = 227.1, \sigma^2 = 265$
(ii) 79

10 (i) Possible answers: cheaper, less time consuming, not all population destroyed if sampling is destructive
(ii) (a) 68.0 to 70.6
90% of random samples give rise to confidence intervals that contain the population mean.
(b) 71.2 is not in the confidence interval so there is a significant difference in the life-span from the national average.

11 (i) $\mu = 4.27, \sigma^2 = 0.00793$
(ii) 4.25 to 4.29
(iii) 9

12 (i) midpoint = 0.145, $n = 600$
(ii) 97.4

13 (i) A random sample is one in which every item in the population has an equal probability of being selected.
(ii) 0.321 to 0.422
(iii) 2240

14 (i) 0.244, 250
(ii) 90%

15 (i) mean = 52.5; variance = 108
(ii) 49.8 to 55.2
(iii) 49

Chapter 5

Exercise 5A (Page 90)

1 (i) 0.266
(ii) 0.826

2 (i) 0.224
(ii) 0.185
(iii) 0.815

3 (i) 0.0580
(ii) 0.0992

4 (i) 25
(ii) 75
(iii) 112
(iv) 288
Assume that mistakes occur randomly, singly, independently and at a constant mean rate.

5 (i) 0.111, 0.244, 0.268, 0.377
(ii) 3

6 (i) 3
(ii) 27.5
(iii) 460 or 461

7 (i) The mean is much greater than the variance therefore X does not have a Poisson distribution.
(ii) Yes because now the values of the mean and variance are similar.
(iii) 0.0117

8 (i) 0.144
(ii) 0.819
(iii) 2.88

9 (i) 0.209
(ii) 0.219
(iii) 6

10 (i) 0.184
(ii) 0.125

? (Page 94)

(i) It is not necessarily the case that a car or lorry passing along the road is a random event. Regular users will change both Poisson parameters, which in turn will affect the solution to the problem.

(ii) With so few vehicles they probably are independent.

(iii) They are more likely in the day than the night. This raises serious doubts about the test associated with this model.

(iv) It could be that their numbers are negligible or it might be assumed they do not damage the cattle grid.

Exercise 5B (Page 95)

1 (i) (a) 0.149
(b) 0.915
(c) 0.0283
(ii) 0.116
(iii) 0.244

2 (i) (a) 0.180
(b) 0.264
(c) 0.916
(ii) 0.296
(iii) 0.549

3 (i) (a) 0.00674
(b) 0.0337
(c) 0.0842
(ii) $T \sim \text{Po}(5.0)$

4 (i) (a) 0.134
(b) 0.848
(ii) 0.0859
(iii) 0.673

5 (i) (a) 0.257
(b) 0.223
(c) 0.168
(ii) 0.340

6 (i) 0.531
(ii) 0.0649
(iii) 0.159

7 (i) 0.0932
(ii) 0.359
(iii) 0.384

8 (i) 0.135
(ii) 0.385
(iii) 0.125

Exercise 5C (Page 102)

1 (i) 0.135 768, 0.140 374, 3.4%
(ii) 0.140 078, 0.140 374, 0.2%
(iii) 0.140 346, 0.140 374, 0.02%

The agreement between the values improves as n increases and p decreases.

2 (i) 0.224
(ii) 0.616

3 (i) 0.104
(ii) 0.560
(iii) 0.762

4 (i) 0.156
(ii) 0.785

5 (i) 0.362
(ii) 0.544
(iii) 0.214
(iv) 0.558

6 (i) 0.0821
(ii) 0.891
(iii) 0.287

7 (i) 0.0788
(ii) 7490

Exercise 5D (Page 107)

1 (i) 0.493
(ii) 0.0730
(iii) 0.687

2 (i) 0.762
(ii) 0.0331
(iii) 0.354

3 (i) 0.188
(ii) 0.991
(iii) (a) 0.963
(b) 1.00
Four lots of 50 is only one of many ways to make 200, so you would expect the probability in part (b) to be higher than that in part (a).

4 (i) (a) 0.614
(b) 0.834
(ii) You must assume that the same number of emails will be received, on average, in the future.
(iii) For longer time periods, there are more and more different ways in which the total can be reached, so the probability increases.

5 (i) (a) 0.617
(b) 0.835
(ii) 0.0592

6 (i) H_0: $\mu = 5.6$ where μ is the mean number of shooting stars
H_1: $\mu < 5.6$
(ii) $X \leqslant 2$ where X is the number of shooting stars
(iii) 0.0824
(iv) The null hypothesis is not rejected. The evidence does not support the astronomer's theory.

7 (i) 0.122
(ii) 0.532
(iii) 0.0135
(iv) 0.229

8 (i) People can be expected to call randomly, independently and at an average uniform rate.
(ii) 0.113
(iii) 0.0209

9 (i) 0.143
(ii) 0.118
(iii) 0.0316

10 (ii) 0 or 1
(iii) 0.0916
(iv) 1 is in the rejection region so there is evidence that the new guitar string lasts longer.

11 (i) 0.0202
(ii) 0.972
(iii) 0.0311

12 (i) There is enough evidence to accept at the 10% significance level that ploughing has increased the number of pieces of metal found.
(ii) There is not enough evidence to accept at the 5% significance level that ploughing has increased the number of pieces of metal found.
(iii) 0.395

13 (i) A Type I error is made if the test finds that the number of white blood cells has decreased when, in fact, the mean number of white blood cells has not decreased; 0.0342
(ii) Accept H_0, there is insufficient evidence to say that the number of white blood cells has decreased.
(iii) 0.915

Chapter 6

? (Page 115)

It is the variance of X.

Exercise 6A (Page 118)

1 (i) (a) $E(X) = 3.1$
(b) $Var(X) = 1.29$

2 (i) (a) $E(X) = 0.7$
(b) $Var(X) = 0.61$

4 (i) 2 (ii) 1 (iii) 9

5 (i) 10.9, 3.09
(ii) 18.4, 111.24

6 (i) $E(2X) = 6$
(ii) $Var(3X) = 6.75$

7 (i) (a) 2.79
(b) 8.97
(c) 20.9

8 (i) $E(N) = 25, Var(N) = 125$
(ii) $E(\overline{N}) = 25, Var(\overline{N}) = 62.5$

Exercise 6B (Page 124)

1 (i) 4, 0.875
(ii) 1.5, 0.167
(iii)

Main course	*Dessert*	*Price*
Fish and chips	Ice cream	\$4
Fish and chips	Apple pie	\$4.50
Fish and chips	Sponge pudding	\$5
Spaghetti	Ice cream	\$4.50
Spaghetti	Apple pie	\$5
Spaghetti	Sponge pudding	\$5.50
Pizza	Ice cream	\$5
Pizza	Apple pie	\$5.50
Pizza	Sponge pudding	\$6
Steak and chips	Ice cream	\$6.50
Steak and chips	Apple pie	\$7
Steak and chips	Sponge pudding	\$7.50

(iv) Mean of $T = 5.5$, variance = 1.042

2 (i) N(90, 25)
(ii) N(10, 25)
(iii) N(−10, 25)

3 0.196

4 (i) 0.0228
(ii) 56.4 minutes
(iii) 0.362

5 (i) 230 g, 10.2 g
(ii) 0.159
(iii) 0.0786

6 (i) N(70, 25)
(ii) N(−10, 25)

7 5.92%

8 (i) 0.266
(ii) No, people do not choose their spouses at random: the heights of a husband and wife may not be independent.

9 0.151

10 (i) 0.0169
(ii) 0.00360

11 Mean = 59.4, standard deviation = 7.09

12 (i) Mean = 3360, variance = 1540 (to 3 s.f.)
(ii) 0.0693

? (Page 130)

With folded paper it is not possible for pieces of paper that are thicker to be offset by others that are thinner, and vice versa.

Exercise 6C (Page 130)

1 N(120, 24)
Assume times are independent and no time is spent on changeovers between vehicles.

2 0.0745

3 0.138

4 0.195

5 (i) N(34, 30)
(ii) N(−4, 30)
(iii) N(0, 30)

6 (i) 0.316
(ii) 0.316

7 (i) N(100, 26)
(ii) N(295, 353)
(iii) N(200, 122)
(iv) N(−65, 377)

8 (i) 0.0827
(ii) 0.310
(iii) 0.5

9 (i) 0.0828
(ii) 0.145
(iii) 0.5
(iv) The situations in 8(i) and 9(i) are the same. 8(ii) considers $3X + 5Y$ whereas 9(ii) considers $X_1 + X_2 + X_3 + Y_1 + \ldots + Y_5$, so the probabilities are different. In both 8(iii) and 9(iii) the mean is zero, so the probability is 0.5, independent of the variance.

10 (i) 312 kg
(ii) Assume that the composition of each crew is selected randomly so that the weights of each of the four individual rowers are independent of each other. This assumption may not be reasonable since there may be some lightweight and some heavyweight crews; also men's and women's crews. If this is so it will cast doubt on the answer to part (i).

11 (i) 0.455
(ii) 93.5 litres

12 0.902
Assume weights of participants are independent since both teams were chosen at random.

13 (i) 0.0374
(ii) 0.238
Assume that no time is lost during baton changeovers and that the runners' times are independent, i.e. that no runners are influenced by the performance of their team mates or competitors. The model does not seem entirely realistic in this.

14 0.195

15 0.350

16 (i) 0.387
(ii) Mean = 10, variance = 11.56
(iii) 0.647

Index